Science and the Enlightenment

CAMBRIDGE HISTORY OF SCIENCE

Physical Science in the Middle Ages
EDWARD GRANT

Man and Nature in the Renaissance
ALLEN G. DEBUS

The Construction of Modern Science:
Mechanisms and Mechanics
RICHARD S. WESTFALL

Science and the Enlightenment
THOMAS L. HANKINS

Biology in the Nineteenth Century:
Problems of Form, Function, and Transformation
WILLIAM COLEMAN

Energy, Force, and Matter: The Conceptual Development of
Nineteenth-Century Physics
P. M. HARMAN

Life Science in the Twentieth Century
GARLAND E. ALLEN

Science and the Enlightenment

THOMAS L. HANKINS

Department of History, University of Washington, Seattle

CAMBRIDGE UNIVERSITY PRESS

CAMBRIDGE

NEW YORK PORT CHESTER

MELBOURNE SYDNEY

For Henry Guerlac and L. Pearce Williams

Published by the Press Syndicate of the University of Cambridge
The Pitt Building, Trumpington Street, Cambridge CB2 1RP
32 East 57th Street, New York, NY 10022, USA
10 Stamford Road, Oakleigh, Melbourne 3166, Australia

First published 1985
Reprinted 1985, 1987, 1988, 1989

Printed in the United States of America

Library of Congress Cataloging in Publication Data

Hankins, Thomas L.

Science and the Enlightenment.

(Cambridge history of science)

Bibliography: p.

Includes index.
1. Science—History. I. Title. II. Series.
Q125.H355 1985 509 84—16988
ISBN 0 521 24349 1 hard covers
ISBN 0 521 28619 0 paperback

Contents

Preface

In 1939 Abraham Wolf published his *History of science, technology and philosophy in the eighteenth century.* Since that time much has changed in the history of science, but no new general study of eighteenth-century science has appeared. The present book is intended to fill the gap – or perhaps a slightly different gap. Wolf's book emphasized technology and instrumentation, whereas mine emphasizes science and ideas. It is impossible to include within the compass of one small volume all of the detail that appeared in Wolf's two large ones. Instead I have outlined the major events in the development of eighteenth-century science, with an eye toward indicating the directions that modern scholarship has taken. In particular, I have attempted to trace the emergence of modern scientific fields. The treatment is not technical, although in some cases (as in the chapter on chemistry), it has been necessary to give an account of actual experiments in order to make the modern interpretations clear.

The history of eighteenth-century science appears in this book as part of the Enlightenment, which means that the viewpoint tends to be French, some might say unremittingly French. My only excuse is that because France was the center of the Enlightenment, my account seemed to flow most naturally from that source, although I would be the first to admit that other viewpoints could be found that would be equally valid. As part of the Cambridge History of Science series, this book is intended primarily for students of the history of science, but I hope that students of the Enlightenment will also find it useful. Wherever possible I have attempted to set the history of science in a broader historical context.

My colleagues in the History Research Group and the History of Science Club at the University of Washington have read portions of the manuscript and have given me valuable criticism. So have Rhoda Rappaport and Jerry Gough, who have helped me with the chapters on natural history and chemistry, respectively. I am also grateful to the series editors and to the reader for the Cambridge University Press, who gave me excellent advice on revising the original draft. John Heilbron and Richard Ziemacki assisted with the illustrations, and Joan Scott and William Scott prepared the line drawings and computer plots of curves.

I also wish to thank the John Simon Guggenheim Foundation and the University of Washington for their support during the time when this manuscript was being written.

CHAPTER I

The Character of the Enlightenment

In 1759 the French mathematician Jean Lerond d'Alembert (1717–83) described a revolution that he saw taking place in natural philosophy:

> Our century is called . . . the century of philosophy par excellence. . . . The discovery and application of a new method of philosophizing, the kind of enthusiasm which accompanies discoveries, a certain exaltation of ideas which the spectacle of the universe produces in us – all these causes have brought about a lively fermentation of minds, spreading through nature in all directions like a river which has burst its dams.[1]

This revolution came to be called the Scientific Revolution, a cultural event associated with great names like those of Galileo Galilei (1564–1642), Johannes Kepler (1571–1630), René Descartes (1596–1650), and Isaac Newton (1642–1727). D'Alembert obviously believed that it was a revolution still in progress in 1759 and one that was continually accelerating. Natural philosophy could never be put back in its former course. As d'Alembert observed, "Once the foundations of a revolution have been laid down it is almost always the succeeding generation which completes that revolution."[2] The seventeenth century had begun the revolution; the eighteenth century would complete it.

The expression "Scientific Revolution" had been coined by mathematicians like d'Alembert, and it was mathematics that appeared to them as the greatest revolutionizing force. In 1700, Bernard le Bovier de Fontenelle (1657–1757), "perpetual secretary" of the Paris Academy of Sciences, first talked about an "almost complete revolution in geometry" that had begun with the analytic geometry of Descartes.[3] Alexis-Claude Clairaut (1713–65), in 1747, attributed the cause of a "great revolution in physics" to Newton's

Principia (1687).[4] In the second half of the eighteenth century, the idea that a Scientific Revolution affecting all aspects of natural science was under way became commonplace, but it is worthwhile remembering that initially the term was used with reference to mathematics and astronomy.

In the preface to his *Histoire* of the Paris Academy of Sciences, Fontenelle argued in 1699 that the new "geometric spirit" could also improve works on politics, morals, literary criticism, even public speaking, and d'Alembert, who shared the anticlerical biases of his fellow philosophes, claimed that if one could smuggle mathematicians into Spain the influence of their clear, rational thinking would spread until it undermined the Inquisition. Thus the philosophers of the eighteenth century believed that the Scientific Revolution was changing all of human activity – not just the natural sciences. "Reason" was the key to a correct method, and the model of reason was mathematics. Of course, "reason" could mean different things. It could mean order imposed on recalcitrant nature, or it could mean common sense (as in the term *reasonableness*), or it could mean logically valid argument, as in mathematics. Because "reason" in any of these meanings was a valuable guide to knowledge and to life, the philosophers of the Enlightenment used it as a rallying cry without worrying too much about its precise definition.

The eighteenth century was called by the French the *siècle des lumières,* the "century of light," because of its emphasis on reason as a path to knowledge. More commonly in English it has been called the Enlightenment. This term, however, was chosen not by a French mathematician but by the German metaphysician Immanuel Kant (1724–1804), who, when asked in 1785 if he believed that he lived in an enlightened age, answered: "No, we are living in an age of enlightenment."[5] Kant, in 1785, like d'Alembert in 1759, thought that the Scientific Revolution was still in progress. Five years after he made his famous statement, the intellectual revolution in France was succeeded by the first great political revolution of modern times. It is paradoxical that a metaphysician should have given the name to an age that so deeply distrusted metaphysics. But the paradox dissolves when we recognize that the Enlightenment was not a fixed set of beliefs but a way of thinking, a critical approach that was supposed to open the way for constructive thought and action. Anyone who believed that man had the power to correct past errors through the use of his reason could find value in the Enlightenment.

Reason and Nature

Throughout the Enlightenment, "reason" was usually extolled in the same breath with "nature," the other key word of the Enlightenment. The connection between these ideas came from seventeenth-century England, where they played a major role in natural theology. Natural theology, in the Middle Ages, had been the domain of those truths that could be found through the use of reason alone, without the Revelation of the Bible. In addition to purely rational arguments, such as the ontological proof of the existence of God, which did not refer to the physical world, there were also truths about God that could be known from an examination of his creation. Such was the argument from design, which stated that the work of a supreme intelligence was evident in the order of the natural world. As the achievements of science grew in the seventeenth century, the argument from design began to replace a priori rational arguments and often even the Revelation of Scripture as the principal evidence for religion. The philosopher John Locke (1632–1704) could enthusiastically claim that "the works of Nature everywhere sufficiently evidence a Deity,"[6] and Robert Boyle (1627–91), the leading scientific experimenter in seventeenth-century England, agreed that he had never seen any "inanimate production of nature, or of chance, whose contrivance was comparable to that of the meanest limb of the despicabilist animal: . . . There is incomparably more art expressed in the structure of a dog's foot than in that of the famous clock at Strasbourg [the most complicated piece of machinery known to the seventeenth century]."[7] If God could be known from his creation, the Bible was not necessary to prove the existence of God.

The motive for this change in the meaning of reason was largely religious, but the implications for science were great. Reason changed from the methods of formal logic to those of the natural sciences, and the laws of reason became identical with the laws of nature. This change in the meaning of reason also caused a change in the way one learned about the natural world. Because the laws of nature were chosen freely by God for his creation, they could be known only by experiment; no logical argument alone could fathom God's free choice. Thus in the seventeenth century experiment became part of the "reasoned" approach to nature.

As long as reason and nature were understood to be manifestations of a divine intellect, they obviously had moral value. The noble Houyhnhnm in Jonathan Swift's *Gulliver's travels* (1726) "thought

1 2 3 4 5 6 7 8 9 10 11 12 Toises.

Fig. 1.1. Two views of nature in the eighteenth century. According to the classical tradition, nature was beautiful only when it was subjected to reason. According to the Romantic view, nature was beautiful when it evoked an emotional or poetic response. Two plates from the *Encyclopédie* illustrate these contrasting views. The first plate accompanies the article on gardening ("Jardinage"). It is a garden by André Lenôtre (1613–1700), who designed the landscaping for the gardens at Versailles and the Tuileries. One sees immediately in his design the influence of the "geometric spirit." The second plate depicts the eruption of Mount Vesuvius in 1754 and illustrates the article on mineralogy ("Minéralogie"). In this Romantic view, nature is beautiful precisely because it is uncontrolled and has escaped the confines of order and reason. Sources: *Encyclopédie, ou dictionnaire raisonné des sciences, des arts, et des métiers,* ed. D. Diderot and J. d'Alembert, 17 vols. of text (1751–65), 11 vols. of plates (1762–77), 4 suppl. vols. of text, 1 suppl. vol. of plates, and 2 suppl. vols. of index (1776–80) (Paris): vol. 1 of plates, "Agriculture, jardinage," pl. III; vol. 6 of plates, "Histoire naturelle volcans, minéralogie," pl. III. By permission of the Syndics of Cambridge University Library.

Nature and Reason were sufficient guides for a reasonable animal, as we pretended to be, in showing us what we ought to do, and what to avoid," and Newton ended his *Opticks* (1704) with the conclusion that by perfecting natural philosophy, "the bounds of Moral Philosophy will also be enlarged" and God's purpose will be exposed "by the light of Nature."[8] The discovery of the laws of nature would necessarily lead to the discovery of God's intentions, which formed the foundation of moral law.

For a deist like Lord Shaftesbury (1621–83) or Voltaire (1694–1778), these arguments from design were adequate to prove the existence of God without any need for the Revelation recorded in the Bible. More daring thinkers such as Spinoza (1632–77) and John Toland (1670–1722) not only dispensed with the Bible but found it reasonable to equate God with nature. Toland invented the term *pantheism* for the belief that God and nature were one and the same. Or, if pantheism were still too theistic (because, after all, according to this philosophy God still existed, in the form of nature), one could follow Denis Diderot (1713–84), Julien Offray de La Mettrie (1709–51), or the Baron d'Holbach (1723–89) and deny the existence of any spiritual God. They insisted that reason and nature were sufficient by themselves.

The Enlightenment was in large part created by this shift from reason as the perfect intelligence to reason as the law of nature. Leaving God out of science may appear from our perspective as an advance in the right direction, at least to the extent that it allowed a more objective scientific method, but it created two paradoxes for the Enlightenment that were never successfully resolved. Both appeared in the attempts of the philosophers of the eighteenth century to create a science of man.*

1. To the extent that the laws of nature were to be discovered by experimentation and observation, they were purely descriptive. They revealed the ordered relations of phenomena and subjected them to rule. They revealed what is, but not what ought to be. On the basis of Newton's universal law of gravitation astronomers could predict the motion of the planets, but they could not say whether gravitation was good or bad. From this point of view it was hopeless to try to extract an ethic from natural science. To live like Swift's Houyhnhmns according to nature and reason would be merely to exist, since nature and reason did not say *how* to live. But the philosophers of the Enlightenment hoped for more from nature. For

*I use the generic term *man,* as in "the science of man," because it was the term used in the eighteenth century. It should be understood as applying to all people, men and women alike.

them the laws of nature and reason contained moral imperatives. Their claim for an objective moral science appeared to contain an implicit contradiction.

2. A second paradox existed in the way natural philosophers of the eighteenth century attempted to apply natural law. Their goal was to find laws governing natural phenomena that would totally and accurately predict future events. The attempt to reach this goal was a search for greater determinism in nature. But at the same time, the laws of nature, when applied to man, were thought to confer ever greater freedom, especially freedom from arbitrary human authority. The inalienable rights of man claimed by the American Declaration of Independence, for instance, were supposed to have been established by "Nature and Nature's God." Yet there seems to be a contradiction in laws of nature that completely determine events in the physical world and at the same time set man free. This is especially true if man is understood to be part of nature. These paradoxes had no easy resolution, and they created a tension in Enlightenment philosophy as long as the philosophers sought virtue or a code of life in natural science.

Science and Literature

There was one way, however, that moral value could be assigned to science without contradiction, and that was through ascribing traditional virtues to the pursuit of science. Fontenelle's eulogies of deceased members of the French Academy of Sciences not only describe the major scientific accomplishments of the age but also extol the purity of the motives for scientific study. In 1688 Fontenelle wrote a treatise on the nature of the eclogue, or pastoral poem. The pastoral convention ascribed simplicity, humility, austerity, want of ambition, and love of nature to the person praised. The same virtues appear in Fontenelle's eulogies of deceased members of the academy. Their purpose in life had been an unselfish search for truth, which in itself was virtuous. In addition to the virtues of the pastoral convention, Fontenelle endowed them with those virtues that Plutarch ascribed to the great men of the Roman world, the Stoic virtues of fortitude, duty, courage, and resolution. Fontenelle used the values expressed in the eclogue and in Plutarch's *Lives* to make the pursuit of natural philosophy virtuous. Although science itself might be entirely objective and without an ethical content, its very objectivity made the natural philosopher a man of virtue. Objectivity was the opposite of self-interest and ambition; the natural philosopher served mankind rather than himself.

Later in the Enlightenment, when the Marquis de Condorcet (1743–94), became secretary of the academy and took over the task of writing the eulogies, he altered their style to make them more polemical. Condorcet had a passion for humanity, a desire to "do good," and a penchant for reform. For him the natural philosopher could no longer remain virtuous in his rustic retreat. The duty to reform society through reason became imperative. Although the imperatives changed, the conclusion remained the same – that the pursuit of natural philosophy was morally good. The most virtuous pursuit of all would be the creation of a science of man that, through reason, would destroy prejudice and superstition and build a new society on objective scientific principles.

The classical and humanistic traditions had placed natural philosophy in the category of letters, or literature. By the nineteenth century the separation of science from literature was almost complete, but the Enlightenment still retained the emphasis on literature. Natural philosophers wanted to be known as "men of letters." All thinkers and workers for the good were members of the "republic of letters." It was a republic because it valued freedom of thought and action and rejected authority. The editors of the great French *Encyclopédie* claimed that their authors composed a society of men of letters. Science during the Enlightenment basically followed a literary ideology in which virtue lay as much in the craft as in the product.

A new critical spirit that questioned all that was not demonstrated had, in fact, appeared in literature before it appeared in natural science. Descriptions of travels, both real and imaginary, allowed writers to establish moral and cultural relativity. Is there any "best" way to live? Pierre Bayle (1647–1706) claimed that a community of atheists could live a completely moral existence, and Montesquieu (1689–1755) used his *Persian letters* (1720) to reveal the absurdity of French customs and institutions. History lost its providential character, and historians strove to present an objective account of past events. In the hands of David Hume (1711–76), history led not to an understanding of God's will but to an understanding of human nature. This critical revolution in literature found strong support from the new and rapidly growing natural philosophy.

Natural philosophy had one advantage over other literary pursuits: it progressed. When the literary world quarreled over whether ancient or modern literature was superior, the moderns inevitably pointed to natural philosophy. In natural philosophy the truths were more obviously cumulative. The seventeenth-century natural philosophers knew what the ancients knew and added to them the truths that had been discovered in more recent times. Drama, poetry, ethics, and jurisprudence did not necessarily progress, because the new might not contain the old. Natural philosophy, on the other hand, had a logical coherence that retained the old as the foundation for the new. A new theorem in geometry did not invalidate old theorems; it built on them and added to them. The "geometric spirit" noted by Fontenelle ensured that the same progress would occur in our knowledge about nature. Thus the natural philosophers appeared to possess a sure method of increasing human knowledge and of improving the human condition.

The ideology of the Enlightenment tended to make natural philosophers into heroes, and in France the greatest hero of all was Newton, partly because he was from England, the source of free thought and liberty in the minds of Frenchmen like Voltaire and Montesquieu, but also because he had solved the riddle of the planets, showing that their motions obeyed the same laws as motions on earth. There was no other problem of such cosmic significance, and therefore no comparable hero. The obvious way for natural philosophy to progress was for natural philosophers to complete Newton's program of research, using his methods. The science of the Enlightenment would then be "Newtonian," and its philosophy would be one of "Newtonianism."

Newton's ideas were circulated in many popular accounts published in England and throughout the Continent. By 1784 there were forty books on Newton in English, seventeen in French, three in German, eleven in Latin, and one each in Portuguese and Italian. Historians of the Enlightenment have quite properly recognized the importance of Newton as a symbol and more recently have shown that his name was even a rallying cry for radical politics and social reform. But this Newtonian ideology existed chiefly at a popular level. When we look at what the natural philosophers themselves were doing, the concept of Newtonianism does not seem to have been very helpful. Newton was, of course, a major source of ideas for further research, but not the only source, and in the laboratory to merely repeat what Newton had done was of little help. Moreover, it was possible to read into Newton's works whatever one hoped to find there. Thus the Marquis de l'Hôpital (1661–1704) made Newton into a supreme rationalist whose laws of motion were a priori deductions of pure thought, whereas the Dutch physicists saw him as a thoroughgoing empiricist.

The case of Nicolas Malebranche (1638–1715) is illustrative. Malebranche, a priest of the Congregation of the Oratory, was also a philosopher, mathematician, and member of the French Academy of Sciences. He had about him a group of mathematicians that included the Marquis de l'Hôpital and Pierre Varignon (1654–1722). This group was responsible for introducing calculus into France, deriving it first from a study of the writings of Leibniz (1646–1716) and later from reading Newton's mathematical papers when they became available. They were also the first in France to accept Newton's law of universal gravitation and the first to appreciate his optical experiments. When Malebranche read the Latin edition of Newton's *Opticks* in 1706, he accepted it with praise, and his friend Varignon was instrumental in publishing the Paris edition (1722) of the French translation.

Malebranche would certainly appear to have been a follower of Newton and therefore a "Newtonian." But he was also regarded as the greatest living disciple of Descartes. Malebranche's famous *Recherche de la Vérité* [*Search for truth*] (1674–5) continued the same philosophy of rationalism emphasized by Descartes. Moreover, it was a theological work. The philosophers of the Enlightenment drew heavily from Malebranche, even as they condemned him as theologian and metaphysician – both bad words during the Enlightenment. Malebranche was at the opposite philosophical extreme from Newton, but he recognized the importance of Newton's achievements before any other Continental figure. Could Newton's earliest supporter on the Continent not be a Newtonian? In circumstances like these, the term *Newtonian* loses all meaning. On the popular level it stands for something, and in the ideology of the Enlightenment it stands for a great deal, but when taken to the laboratory and to the mathematician's desk it is too general and imprecise to be of much help to the historian of science.

The Categories of Science

In our search for adequate categories of scientific thought during the Enlightenment, we might assume that we could follow the lines of division suggested by the modern scientific disciplines, but eighteenth-century science was not organized in terms of the modern disciplines. For example, we will go wrong if we look for the "origins" of modern physics in the eighteenth century, both because such a search assumes that there was a field in the eighteenth century that dealt with the same phenomena as modern physics and because the terms that are used in physics have changed their meanings over time. In fact, "physics," at the beginning of the Enlightenment, was "the science that teaches us the reasons and causes of all the effects that Nature produces," and this included both living and nonliving phenomena. Medicine and physiology, as well as the study of heat and magnetism, were part of physics. In the seventeenth century the physician and the physicist were the same person. Moreover much of what we would now call physics was called "mixed mathematics" in the eighteenth century. Mixed mathematics included astronomy, optics, statics, hydraulics, gnomonics (concerned with sundials), geography, horology (concerned with clocks), navigation, surveying, and fortification (see Figure 1.2).

The same difficulties arise in all parts of science. Chemistry was practiced largely by medical doctors, who saw it as part of their field. Because it included the study of the mineral kingdom, chemistry overlapped with natural history, the science that described and classified all forms of nature (see Chapter V). Chemistry also blended indistinguishably into physics, because the study of heat and the gaseous state were part of chemistry. Our modern sciences of zoology, botany, geology, and meteorology were all subsumed (at least in part) under natural history. The names *zoology, botany, geology,* and *meteorology,* which had been used earlier with slightly different meanings, were familiar, but both *biology* and *sociology* were names and fields that were created in the nineteenth century.

During the eighteenth century all of these categories began to shift into the arrangements that are familiar to us today, but it was a gradual process. The creation of the new scientific disciplines was probably the most important contribution of the Enlightenment to the modernization of science, and one that we might easily overlook. It marks the Enlightenment as a period of transition between the old and the new.

The changing categories of science during the Enlightenment were a reflection of changing views of nature and its study. What we call *science* today was more commonly called *natural philosophy* during the Enlightenment. (It should not be confused with natural *history.* The editors of the *Encyclopédie* (see Chapter VI) ascribed natural history to the mental faculty of memory, and natural philosophy to the mental faculty of reason.) Natural philosophy was still part of philosophy and still struggled with philosophical questions such as those concerning the existence of the soul, the activity and passivity of matter, the freedom of the will, and the existence of God. Today we regard such questions as extraneous to natural science, but this was not the case in the Enlightenment. Even in rational mechanics (the mathematical study of motions and the forces producing them), Pierre-Louis-Moreau de Maupertuis (1698–1759) claimed that his "principle of least action" proved the existence of God, and Madame

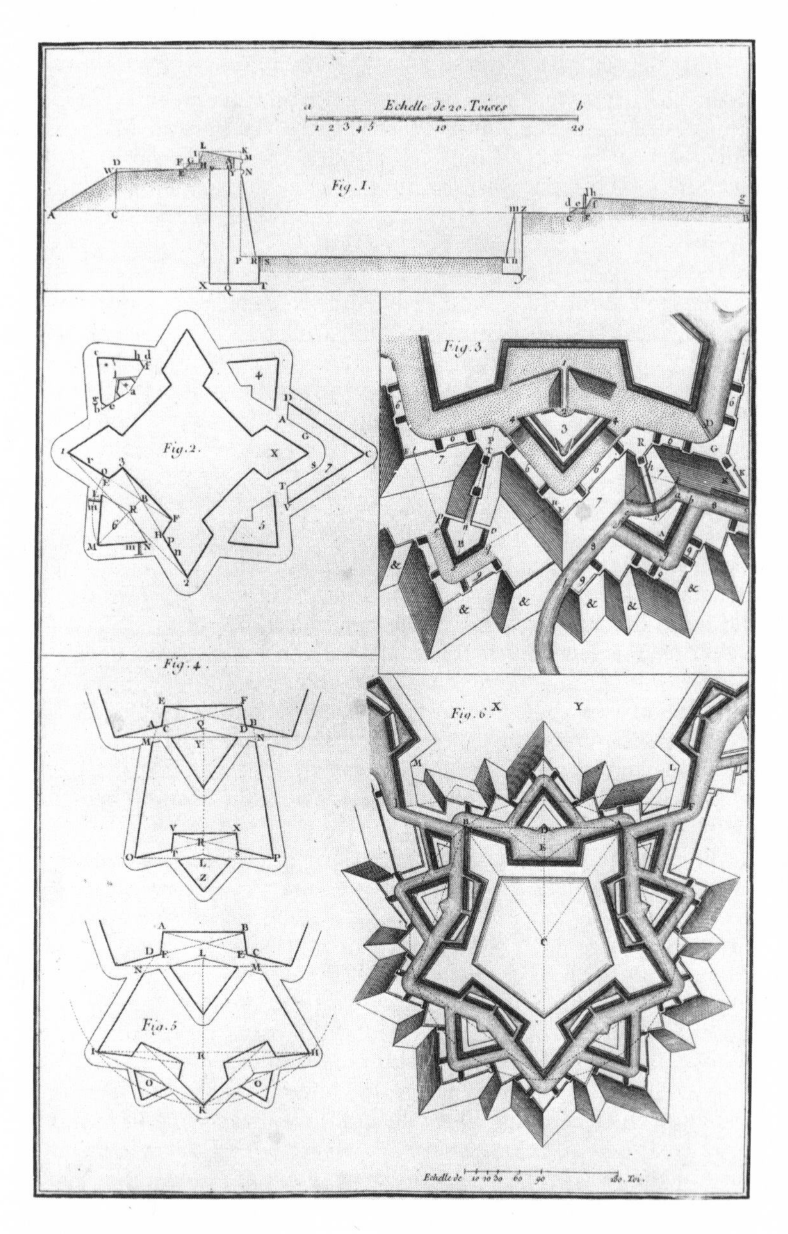
Echelle de 20. Toises
Fig. I.
Fig. 2.
Fig. 3.
Fig. 4.
Fig. 5
Fig. 6.

du Châtelet (1706–49) supported the Leibnizian theory of mechanics because it gave a better accounting of free will. To understand the science of the Enlightenment, we must enter the eighteenth-century scientist's laboratory without too many preconceptions.

The Mechanical Philosophy

In general, the philosophers of the Enlightenment accepted the mechanical philosophy that they had inherited from their seventeenth-century forebears. These predecessors had succeeded in removing from natural philosophy the concept of final causes and most of the Aristotelian concepts of form, substance, and accident that had dominated medieval thought. The mechanical philosophy required that the changes observed in the natural world be explained only in terms of motion and rearrangements of the parts of matter. But beyond these basic assumptions, the mechanical philosophers were divided. They differed primarily in their theories concerning the causes of motion and change. Was matter moved by an external power, an internal power, or by no power at all? All three of these positions had their supporters, and all three of them suggested ideological positions that extended beyond natural philosophy itself.

Fig. 1.2. Fortification. The study of fortification was part of "mixed mathematics" during the Enlightenment. The best places to obtain an education in mathematics were the military schools, most notably the École Royale de Génie at Mézières. In this plate from the *Encyclopédie,* Figure 1 is a cross-section of a fortification. To approach the stronghold, the attacking army must run up the slope (*gh*) into defensive fire from the parapet (*ch*), then cross the ditch and scale the escarpment (*RN*) to reach the parapet (*LMN*). Figures 2 through 6 are views of the fortification from above. The purpose of the design is to maximize the cross fire against an assaulting army, leaving no walls unprotected and no easy target for the enemy's siege guns. Because the effective range of a musket was short, the design required frequent bastions and "counterposts" (freestanding bastions) to protect the walls of the main fortification. The fortification depicted here is the most advanced design of the great French military architect Sébastien Le Prestre, Marquis de Vauban (1632–1707). *Sources: Encyclopédie, ou dictionnaire raisonné des sciences, des arts, et des métiers,* ed. D. Diderot and J. d'Alembert, 17 vols. of text (1751–65), 11 vols. of plates (1762–77), 4 suppl. vols. of text, 1 suppl. vol. of plates, and 2 suppl. vols. of index (1776–80) (Paris): vol. 1 of plates, "Art militaire, fortification," pl. IV. By permission of the Syndics of Cambridge University Library.

Descartes, one of the originators of the mechanical philosophy, believed that there were no forces or powers in matter. God, he held, had created the universe as a perfect clockwork mechanism that was able to function thereafter without any intervention. Motion, Descartes argued, is communicated from one part of matter to another by direct contact. All that we observe in the world is matter in motion, and the concept of matter in motion is sufficient to explain all phenomena.

Malebranche carried this denial of the existence of force to an extreme. Any force or power in matter, he claimed, would be power removed from God; he therefore denied that one piece of matter could ever in any way affect another. When two bodies collide, Malebranche argued, no force is exerted between them. The collision is merely an "occasion" for God to act, and since there can be no activity in matter, it is God who changes the motions of bodies. The "occasional cause" was no real cause at all but merely an opportunity for God, the source of all power, to act. In support of his position, Malebranche pointed out that we never see causes: All that we actually observe are the changing motions of matter. We assign causes to these changes in order to interrelate natural events in an orderly way. The British philosophers George Berkeley (1685–1753) and David Hume elaborated this theme into a more complete critique of all theories of causality and emphasized the empirical and probabilistic nature of knowledge about the physical world.

Newton accepted the mechanical philosophy, but not the denial of the existence of force. Although he did not claim to know the nature of the forces of gravitation, cohesion, elasticity, and the like, he was convinced of their existence and built his mechanics from the assumption that matter consists of inert particles with a force of attraction or repulsion acting between every pair of particles. He denied the existence of any innate forces or powers, with the possible exception of a force of inertia or passivity, which he could not explain by external forces. In his alchemical and religious studies he also speculated on the possibility of an all-intrusive power or spirit but concluded that such a power was imposed on matter and not internal to it.

Leibniz argued that force was internal to matter. In fact, the force was more "real" than the matter in which it was contained, because matter, for Leibniz, was only a phenomenon, a sensible manifestation of the relationship between the active metaphysical substances of which the universe is composed. If one divides matter finely enough, one must come to something that is no longer matter, and these nonmaterial metaphysical entities are the sources of the power and direction that we observe in the world.

These versions of the mechanical philosophy in the seventeenth century were prompted in large part by religious motives. Descartes was fighting against the Renaissance animism that described nature as a living thing and recognized souls in all of its parts. Newton was not an animist, but he believed that to deny forces in nature was to deny God. A clockwork universe that did not need to be repaired could get along without the clockmaker. For Newton, the forces exerted between the parts of matter were actions either of God or of his agents. Finally, Leibniz believed that only by placing the attributes of soul, power, and will within the substances of nature can we explain the workings of nature and of God's will.

Each of these arguments regarding the place of force in nature responded to a religious motive, but in each case the argument could be turned around. Malebranche's criticism of the concept of force was valid whether one believed in God or not. Newton's forces of attraction and repulsion acting on the parts of matter could be ascribed to the machine itself and did not have to be ascribed to the actions of a God. By placing action within matter, Leibniz invited philosophers to identify God with nature, or, one might say, to attribute power to nature and forget about God. The skeptics during the Enlightenment were able to skillfully reverse the religious arguments of the creators of the mechanical philosophy.

Science and Philosophy

Many of the scientists of the Enlightenment would have liked to have declared such disputes a waste of time and to have banished them from science entirely, but they could not. As long as one accepted the mechanical philosophy, physical theory had to respond to these debates. Can one dispense with forces in the science of mechanics, as d'Alembert and Lazare Carnot (1753–1823) attempted to do, or must one join Johann Bernoulli (1667–1748) and Leonhard Euler (1707–83) in accepting them? If we do accept a concept of "force," is it proportional to velocity (as theorized by the later Cartesians), to the change in the velocity (following Newton), or to the square of the velocity (following Leibniz)? In order for the concept of force to be helpful in a quantitative science like mechanics, it is necessary to know how to measure force.

The same problems appear in all the sciences. Are living things merely machines, or do they contain a vital principle that distinguishes them from nonliving things? Can chemical phenomena be explained by the laws of motion, or is there a power in nature that cannot be reduced to mechanical action? These questions are as much a part of science as they are a part of philosophy. The debate over the nature of force remained at the intersection between natural science and philosophy, as it had been in the seventeenth century.

It is significant that the chief expositors of these philosophical positions – Descartes, Newton, and Leibniz – were also the leading mathematicians of the Scientific Revolution. Each argued that his method for obtaining knowledge went beyond the methods of mathematics, but each also saw the rational methods of mathematics as the essential core and model for his thought. Mathematics set the style for the science of the Enlightenment. Even philosophers like Diderot and Comte de Buffon (1707–88), who during the 1750s accused their fellow scientists of an excessive reliance on mathematics, did not repudiate the application of reason, which they believed to be best exemplified in mathematics. This identification of natural law with reason gave the philosophers of the eighteenth century an extremely optimistic view of the possibilities of the new science. D'Alembert claimed at mid-century that nothing now stood in the way of the philosophers except bigotry and superstition. Once the proper scientific method was recognized and applied, a steady enlargement of human knowledge and a steady improvement of the human condition would be the inevitable result.

CHAPTER II

Mathematics and the Exact Sciences

The triumph of mathematics during the Enlightenment can be judged by the strangely conflicting testimony of those mathematicians who helped to create it. In a famous letter of September 21, 1781, Joseph-Louis Lagrange (1736–1813) wrote to his mentor Jean d'Alembert that he feared mathematics had reached its limit. He compared mathematics to a mine whose precious minerals had been pursued deeper and deeper into the earth to the limit of human accessibility. "Unless new seams of ore are discovered, it will be necessary to abandon it sooner or later."[1] Bernard Fontenelle had sounded the same warning as early as 1699, and Diderot used the exhaustion of mathematics as the best argument for turning to the more descriptive sciences of natural history, anatomy, chemistry, and experimental physics. He argued that like the pyramids of Egypt, the creations of mathematicians would stand for centuries but that like the pyramids, they could have little added to them and little practical use could be made of them. D'Alembert and the Marquis de Condorcet, on the other hand, urged mathematicians to keep the faith and trust to the future, even though the future for mathematics was uncertain.

The Meaning of Analysis

One wonders why the mathematicians of the eighteenth century, who had witnessed the spectacular success of their discipline during their lifetimes and who had seen mathematics become the prime exemplar of reasoned thought and the model against which the other sciences were to be judged, were so uncertain about its future. In 1810, Sylvestre-François Lacroix (1765–1843) gave a clue to the source of this pessimism when he wrote that "the power of our

PRINCIPIA MATHEMATICA. 39

nil impediret, recta pergeret ad *c*, (per leg. I.) describens lineam *Bc* æqualem ipsi *AB*; adeo ut radiis *AS*, *BS*, *cS* ad centrum actis, confectæ forent æquales areæ *ASB*, *BSc*. Verum ubi corpus venit ad *B*, agat vis centripeta impulsu unico sed magno, efficiatque ut corpus de recta *Bc* declinet & pergat in recta *BC*. Ipsi *BS* parallela agatur *cC*, occurrens *BC* in *C*; & completa secunda temporis parte, corpus (per legum corol. I.) reperietur in *C*, in eodem plano cum triangulo *ASB*. Junge *SC*; & triangulum *SBC*, ob parallelas *SB*, *Cc*, æquale erit triangulo *SBc*, atque ideo etiam triangulo *SAB*. Simili argumento si vis centripeta successive agat in *C*, *D*, *E*, &c. faciens ut corpus singulis temporis particulis singulas describat rectas *CD*, *DE*, *EF*, &c. jacebunt hæ omnes in eodem plano; & triangulum *SCD* triangulo *SBC*, & *SDE* ipsi *SCD*, & *SEF* ipsi *SDE* æquale erit. Æqualibus igitur tempori-

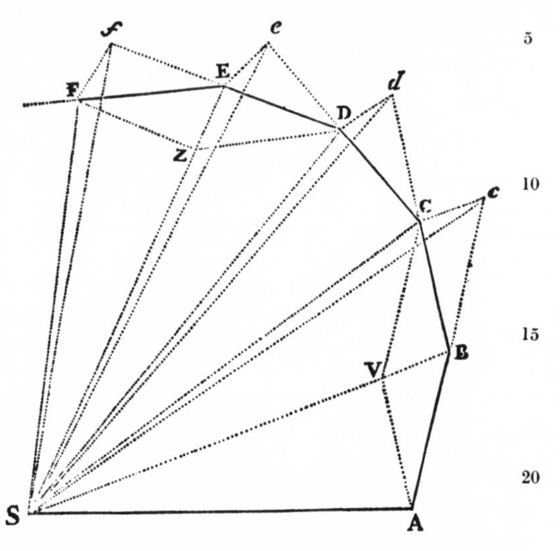

LIBER PRIMU

5

10

15

20

25

1] (per leg. I.) M [IN]

4–22 fig.] Z V FZ ZD VA *om.* E_1 *but add.* E_1i *and* Z V FZ *add.* E_1a

7–8] efficiatque: faciatque *M* E_1 *but* $E_1i = E_3$

8] ut *om. M* E_1 *but add.* E_1i E_1a *and MS Errata to* E_1a | de: a *M* E_1 *but* $E_1i = E_3$

9] declinet: deflectere *M* E_1 *but changed in* E_1a *and MS Errata to* E_1a *to* deflectat *and* $E_1i = E_3$

10] pergat: pergere *M* E_1 *but* E_1i E_1a *and MS Errata to* $E_1a = E_3$

16–17] (per legum corol. I.) *M* [*IN*]

26] omnes *om. M* E_1 *but add.* E_1i E_1a

Fig. 2.1. The transition from geometry to analysis. Although Newton was one of the discoverers of calculus, he did not use it in his *Philosophiae naturalis principia mathematica* [*Mathematical principles of natural philosophy*] (1687). The page from this work reproduced in Figure 2.1 (*left*) shows Newton's proof of Kepler's equal-area law of planetary motion; the proof proceeded entirely by geometry or "synthesis." Compare the page from Joseph-Louis Lagrange's *Mécanique analytique* [*Analytical mechanics*] (1788),

9. De cette maniere la formule générale du mouvement $\Gamma + \Delta = 0$ (art. 2) ſera transformée en celle-ci,

$$\Xi\,\delta\xi + \Psi\,\delta\psi + \Phi\,\delta\varphi + \&c = 0,$$

dans laquelle on aura

$$\Xi = d.\frac{\delta T}{\delta d\xi} - \frac{\delta T}{\delta\xi} + \frac{\delta V}{\delta\xi}$$

$$\Psi = d.\frac{\delta T}{\delta d\psi} - \frac{\delta T}{\delta\psi} + \frac{\delta V}{\delta\psi}$$

$$\Phi = d.\frac{\delta T}{\delta d\varphi} - \frac{\delta T}{\delta\varphi} + \frac{\delta V}{\delta\varphi}$$

&c,

en ſuppoſant

$$T = S\left(\frac{dx^2 + dy^2 + dz^2}{2dt^2}\right)m,\ V = S\Pi m,$$

$$\&\ d\Pi = P\,dp + Q\,dq + R\,dr + \&c.$$

Si donc dans le choix des nouvelles variables ξ, ψ, φ, &c, on a eu égard aux équations de condition données par la nature du ſyſtême propoſé, enſorte que ces variables ſoient maintenant tout-à-fait indépendantes les unes des autres, & que par conſéquent leurs variations $\delta\xi$, $\delta\psi$, $\delta\varphi$, &c, demeurent abſolument indéterminées, on aura ſur le champ les équations particulieres $\Xi = 0$, $\Psi = 0$, $\Phi = 0$, &c, leſquelles ſerviront à déterminer le mouvement du ſyſtême; puiſque ces équations ſont en même nombre que les variables ξ, ψ, φ, &c, d'où dépend la poſition du ſyſtême à chaque inſtant.

Mais quoiqu'on puiſſe toujours ramener la queſtion à cet état, puiſqu'il ne s'agit que d'éliminer par les équations de condition, autant de variables qu'elles permettent de le faire, & de prendre enſuite pour ξ, ψ, φ, &c, les variables

containing "Lagrange's equations," in which the victory of analysis over synthesis is complete (Figure 2.1, *right*). In Lagrange's mechanics these equations take the place of Newton's three laws of motion. *Sources:* Isaac Newton, *Philosophiae naturalis principia mathematica* (London, 1687), sec. II, proposition I; and Joseph-Louis Lagrange, *Mécanique analytique* (Paris, 1788), vol. I, pt. 2, "La dynamique," sec. 4, pars. 9–10. By permission of the Syndics of Cambridge University Library.

analysis is practically exhausted."[2] What seemed to be exhausted was "analysis," that particular field of mathematics that had dominated the eighteenth century.

The eighteenth century defined analysis as the method of resolving mathematical problems by reducing them to equations. Thus analysis included algebra, but more especially it employed differential and integral calculus and their applications in mechanics. Descartes, Newton, and Leibniz had discovered this new field of mathematics, and the mathematicians of the eighteenth century exploited it with spectacular results, but Lagrange was essentially correct. Without the infusion of new ideas from other heretofore undiscovered fields of mathematics, "analysis," as they understood it, had nearly run its course. The very success of analysis during the Enlightenment was the reason for pessimism about any future progress.

The word *analysis* was also used in the eighteenth century to describe the proper scientific method. Newton had advocated a method of analysis and synthesis in experiment whereby complex phenomena were to be analyzed into simple components (as when he decomposed white light into colored rays) and then recombined by "synthesis" (as when the colored rays were focused back together by a lens to obtain white light again). The distinction between analysis and synthesis was as old as Aristotle, who had used it to distinguish among different approaches to the solution of mathematical problems. Philosophers had often assumed that the methods of mathematics had their parallels in the reasoning employed in natural philosophy, but these parallels were made most explicit by Newton's teacher Isaac Barrow (1630–77), who stated in 1665 that "*Analysis* . . . seems to belong no more to *Mathematics* than to *Physics, Ethics* or any other Science. For this is only . . . a certain Manner of using Reason in the Solution of Questions, and the Invention or Probation of Conclusions, which is often made use of in all other Sciences."[3]

Throughout the seventeenth century, analysis and synthesis were regarded as two separate methods: Analysis, or "resolution," was a method of discovery, whereas synthesis, or "composition," was a method of proof. With Newton, however, they became one method and were applied not merely to the course of thought but also to the actual doing of experiments. Analysis, for Newton, consisted in "making experiments and observations and in drawing general Conclusions from them by Induction."[4] Analysis and synthesis meant the resolution and composition of nature to understand how it operated. At the end of the eighteenth century, Antoine Lavoisier

(1743–94) invoked the method of analysis and synthesis in his *Traité de chimie* [*Treatise on chemistry*] in which he described an experiment that allowed him to resolve the atmosphere into its constituents and then bring them back together again to create an "air" indistinguishable from atmospheric air.

The major philosopher of method in France, Etienne Bonnot, Abbé de Condillac (1714–80), carried the eighteenth-century enthusiasm for analysis to an extreme. He insisted that all discovery must proceed by analysis, and that synthesis, which had been an important part of Newton's method, could be employed only to demonstrate the validity of a mathematical proposition or experimental conjecture after the answer was known. Thus he believed that the sciences, including mathematics, should be taught by analysis, since the proper way to learn was to duplicate the process of discovery.

Voltaire shared Condillac's enthusiasm for analysis and argued that "the only way man can reason on the objects [of experience] is analysis." He insisted that Newton had stopped "whenever this torch was lacking to him."[5] The greatest errors of the ancients, according to Condillac, had been their adoption of "synthesis." The theorems of Euclidean geometry did not contain unknowns. The proposition to be proved preceded the proof. Aristotle's philosophy was also a demonstration of principles already known or assumed. Condillac claimed that whatever success Euclid and Aristotle had had was due to their concealed use of analysis. With the coming of the new philosophy, the true value of analysis had been revealed and had made possible the great advances of the Scientific Revolution. Those with more mathematical knowledge than Condillac were not prepared to accept his denigration of geometry, but his enthusiasm for analysis was characteristic of the Enlightenment.

The shift of mathematical emphasis from geometry to analysis was brought about largely by efforts to solve problems in mechanics. Calculus had been created to deal with the problem of motion, and the new mathematical techniques discovered in the eighteenth century were all responses to the challenges of mechanics. In no other century was mathematics so closely related to physical problems, and in no other century was mathematics so narrowly focused on the calculus of "analysis." The physical problems to which calculus was applied were not usually concerned with practical mechanics, which was still the business of artisans. Instead mathematicians pursued "rational mechanics," in which the physical object was reduced to a few idealized properties that were capable of quantification. This is not to say that mathematicians had no use for

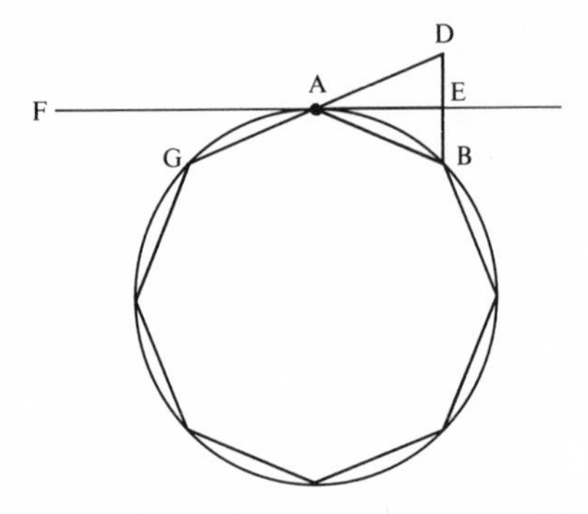

Fig. 2.2. The "polygon curve" and the "rigorous curve." A ball moving along the "polygon curve" would follow the line *AD* if released at *A*. A ball moving along the "rigorous curve" would follow the tangent *AE*. The displacement back onto the circle is either *DB* or *EB*, depending on which curve is assumed. But DB is always twice *EB*, even when the polygon curve is made to approximate the rigorous curve ever more closely. D'Alembert attempted to resolve the paradox by saying that in the case of the polygon curve the centripetal force should be considered as a series of discrete blows, one at each vertex of the polygon, whereas in the case of the rigorous curve the centripetal force should be considered as a continuous push. The displacement *DB* from the tangent to the polygon curve would then be made by uniform motion, and the displacement *EB* from the rigorous tangent would be made by accelerated motion. With this understanding, it turned out that the displacements *DB* and *EB* would be covered in the same length of time.

practical problems. Leonhard Euler, the greatest analyst of the Enlightenment, created mathematical theories to predict the buckling of columns and beams; optimal designs for ships' hulls, sails, and anchors; a theory of anachromatic lenses; theories to describe the motions of vibrating strings and metal plates; designs for waterwheels and turbines; and a host of other applications, but the emphasis remained mathematical and theoretical. Only in astronomy did the new analysis show immediate practical results in the increased precision of astronomical tables and in the creation of new theories concerning the shape and motions of the earth and other heavenly bodies.

Motion Along a Curve

The problem of drawing tangents to a curve provides a good illustration of the close connection between mathematics and mechanics in the eighteenth century. This is a mathematical problem, but if the curve is taken to be the trajectory of a mass point, then the tangent is also the direction of motion of the particle at the point of tangency, and the mathematical problem becomes a physical problem. Leibniz, in his differential calculus, broke up the curve into many little straight lines, creating a "polygon curve." As the number of sides of the polygon is increased, the polygon approximates the curve ever more closely. He defined the tangent at any point on the curve as the extension of the side of the polygon whose vertex is at that point.

When this method of drawing the tangent was used to find the law of centripetal force, however, it created a paradox that exercised the minds of most of the early creators and extenders of differential calculus, including Leibniz himself, Pierre Varignon, and the Marquis de l'Hôpital in France; Jakob Hermann (1678–1733) in Switzerland; and Johann Bernoulli in Holland. The paradox came about because the centripetal force was measured by the distance that the moving body had to be deflected from the tangent in order to keep it on the curved trajectory. In Figure 2.2, *FE* is the "rigorous" tangent to the curve at *A* (derived by classical geometric techniques), and the line *GD* is the tangent as defined by Leibniz. As the number of sides of the polygon is increased, Leibniz's tangent *GD* comes ever closer to the geometric tangent *FE,* and in the limit the two lines coincide. The centripetal force is measured by the displacement from the tangent back to the curve. This displacement will be *DB* from the Leibnizian tangent and *EB* from the geometric tangent. But these two measurements of the displacement from the tangent do not approach each other when the number of sides of the polygon is increased. In fact the displacement *DB* is always twice *EB*, no matter how many sides the polygon has and no matter how small the angle becomes. No wonder that an embarrassing factor of two frequently separated the values that mathematicians obtained for the law of centripetal force.

The displacement *DB* or *EB* not only measured the centripetal force. It was also the "second difference" in calculus – the quantity that Leibniz wrote as d^2x. Depending on whether one calculated this "second difference" from the "polygon curve" (*DB*) or from the "rigorous curve" (*EB*), the results differed by a factor of two. For fifty years mathematicians worked to remove this paradox from calculus. D'Alembert was still trying to resolve it in his articles for the *Encyclopédie* in 1751, and the Chevalier de Louville (1671–1732) actually proposed an experiment in 1732 to the French Academy of Sciences to see if a ball released from a circular orbit would fly away by the Euclidean tangent or by some other line that was the extension of the side of the polygon curve. Most mathematicians were not as naive as Louville, but the paradox remained until the derivative replaced the differential as the fundamental concept of calculus. Only by resolving the mathematical problem of how to define the tangent in calculus could the physical problem of centrifugal force be adequately treated.

The difficulties of handling infinitesimal quantities, as seen in the debate over the polygon curve, were reflected in the physical problem of hard bodies. Are the particles of which matter is composed absolutely hard, as Newton and the atomists said, or are they "elastic" (a new word in the late seventeenth century)? If they are absolutely hard, then objects that collide must instantaneously change their velocities from one value to another without passing through intermediate velocities. On the other hand, "elastic" bodies would be deformed and would rebound when they collide, so that changes in their velocities would be continuous. The polygon curve simulated hard-body impact, the moving object being kept on its trajectory by a blow at each vertex, whereas between blows it moved in a straight line, following the law of inertia. The "rigorous curve" was like an elastic impact in which all changes of velocity were continuous. Each of these kinds of interactions – discrete impacts and continuously applied forces – could be thought of in terms of the other: A continuously applied force could be understood as a rain of tiny impacts, and a single blow could be understood as a very rapid acceleration. But the two interpretations of force were not the same even in the limiting case, and this caused trouble for mathematicians.

From our perspective, it is surprising that mathematicians in the eighteenth century generally felt more comfortable thinking in terms of discrete blows rather than uniformly applied forces. Descartes's mechanics had been based on impact, but those mathematicians of the eighteenth century who accepted the idea of perfectly hard bodies were not necessarily disciples of Descartes. D'Alembert, Maupertuis, Colin Maclaurin (1698–1746), and Lazare Carnot all accepted the concept of perfectly hard bodies, whereas Leibniz, Johann Bernoulli, and Euler denied the existence of any such objects. D'Alembert tried to resolve the dilemma by doing away with forces altogether and by building a mechanics based on motions alone. Euler and Johann Bernoulli recognized that some dynamic concept such as that of force was necessary in mechanics. They denied the existence of perfectly hard bodies and insisted that all changes of motion must be continuous, but they tended not to involve themselves in the metaphysical complexities of the subject. Leibniz's solution to the problem was not only to deny the existence of perfectly hard bodies but to go further and deny that matter had any real meaning except as the manifestation of force. In these very fundamental questions, mathematics and mechanics were inextricably bound up with philosophical debates throughout the Enlightenment.

Mathematics and mechanics also advanced through the study of curves defined by some mechanical motion. The most famous of these curves was the cycloid, already studied by Pascal (1623–62) and Huygens (1629–95) in the seventeenth century. The cycloid is the curve traced by a spot on a rolling wheel. Other types of curves that were studied were the catenary (the shape of a chain suspended between two fixed points); the brachistachrone (the path by which an object slides from one point to another that is not on the same vertical line in the shortest possible time); the involute (the curve traced by the end of a string as it is unwrapped from another curve); and the tractrix (the path of a body that is dragged over a resisting horizontal surface by a cord of which one end moves along a straight line).

Some of these problems allowed mathematicians to perfect known methods. Others led to completely new fields of analysis. The calculus of variations is the problem of finding a curve or path that maximizes or minimizes some property. It began with Newton's attempt to discover the shape of a solid that would offer the least resistance to a fluid. It continued in 1696 with Johann Bernoulli's solution to the problem of the brachistochrone (it turned out that the correct curve was the cycloid) and was greatly furthered by both the Bernoulli brothers, Johann and Jakob (1654–1705), in their study of isoperimeters (the problem of finding the shape of a

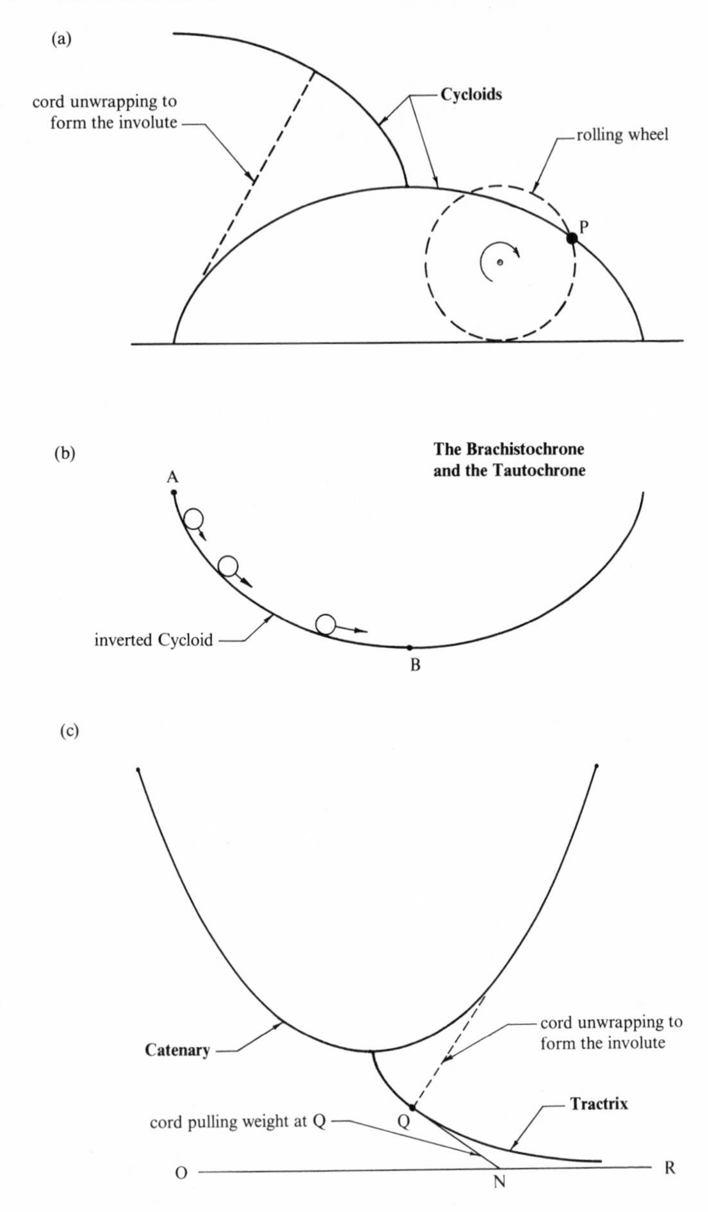
(a)
cord unwrapping to
form the involute
Cycloids
rolling wheel
P
(b)
The Brachistochrone
and the Tautochrone
A
inverted Cycloid
B
(c)
Catenary
cord unwrapping to
form the involute
Tractrix
cord pulling weight at Q
Q
O
N
R

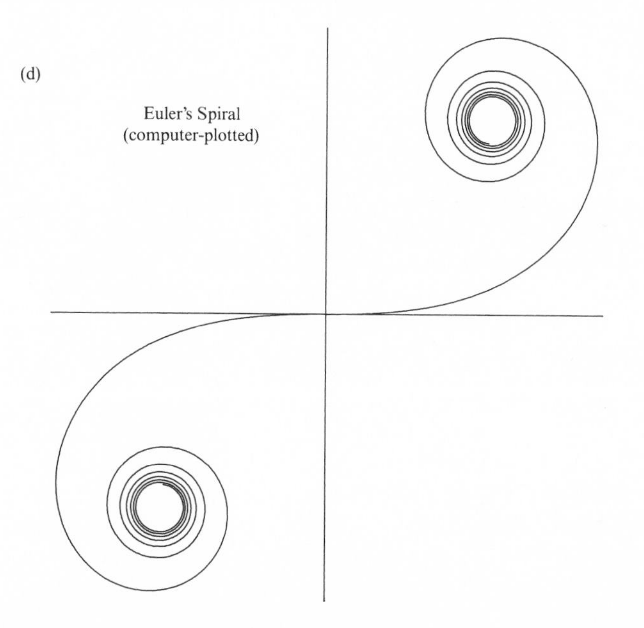

Fig. 2.3. Curves. The cycloid (a) is the path traced by a point (*P*) on the rim of a rolling wheel. The brachistochrone (b) is the curve along which a body slides from one point (*A*) to another point (*B*) under the action of gravity in the shortest time. The tautochrone is the curve along which a body arrives at a given final point in the same time no matter from where on the curve it is started.

The brachistochrone and the tautochrone are both inverted cycloids. The involute of a given curve is another curve traced by the end of a string unwrapped from the given curve. The involute (a) of the cycloid is also a cycloid.

The catenary (c) is the form assumed by a chain suspended by its ends. The tractrix is the path of a body (*Q*) that is dragged over a resisting horizontal surface by a cord of which one end (*N*) moves along a straight line (*OR*). The involute of the catenary is the tractrix.

Euler's spiral (d) is the curve of a watch spring; its radius of curvature is inversely proportional to the arc length of the curve measured from some point of reference. Euler's spiral, to the author's eye, is the most aesthetically pleasing of all the curves. It has applications in Fresnel's theory of the diffraction of light and in the design of railway lines, where it is often used as the transition curve.

surface of maximum area for a perimeter of given length). The methods for solving problems of this kind were generalized by Euler in his *Methodus inveniendi lineas curvas maximi minimive proprietate gaudentes* [*The art of finding curved lines that enjoy some maximum or minimum property*] (1744) and further extended by Lagrange in the 1750s and in his *Mécanique analytique* [*Analytical mechanics*] of 1788.

Attempts to describe the motion of a vibrating string also provided new opportunities for mathematicians. In 1746, d'Alembert first found and solved the wave equation, which is the general solution for describing the motion of a vibrating string. The problem required the use of partial differential equations – that is, equations in differential calculus that have more than one variable. It also raised questions about the nature of mathematical functions. Euler perfected this theory too, and he and d'Alembert became embroiled in a debate over the definition of mathematical functions. D'Alembert believed that any mathematical curve, if it were to be treated adequately in calculus, had to be like a vibrating string: continuous, without any breaks or kinks in it, and stretched between fixed end points. Euler argued that these were unnecessary restrictions and that a mathematical function could describe any curve, even one "drawn by hand," as long as the function was defined over the periodic interval. In this particular debate the controversy rapidly left the physical problem behind and turned on fundamental points in the theory of functions.

Mechanical Principles

Throughout the century, analysis was used to attack ever more difficult mechanical problems. The motion of an extended rigid body such as a gyroscope or top; the flow of water in differently shaped tubes and containers; the vibrations of elastic surfaces and solids, and the motions of complex systems of masses joined by rods, strings, or other kinds of constraints – all of these complicated motions were conquered by analysis. Analysis required only pencil and paper. The only machine employed by rational mechanics was the mind, leading many to question, as did Diderot, the applicability of these splendid theories to the design of real machines that had to pump water, hold up bridges, and produce textiles. That application had to wait until the nineteenth century and the creation of the "mathematical physicist," a scientist equally at home in mathematics and in the laboratory.

Newton had set rational mechanics on the course that it followed during the Enlightenment, but there was more to mechanics than simply the sophisticated use of Newton's laws of motion. Newton's laws were not adequate to deal with all of the mechanical phenomena studied during the Enlightenment. His laws described the motions of individual mass points subjected to forces such as the force of gravitation. They were fine for describing the motions of planets, but they did not describe what would happen to a rigid body composed of many mass points held in a rigid configuration, nor could they describe the motion of a fluid or of the waves in a stretched string. New principles and new concepts had to be developed to deal with these other motions.

These problems became the particular subject of the mathematicians from Basel, Switzerland. Jakob Bernoulli was the first to exploit Leibniz's calculus in mechanics, and he was followed by his brother Johann, his nephew Daniel (1700–82), and an illustrious train of Bernoulli descendants throughout the century. Other Basel mathematicians were Jakob's pupil Jakob Hermann and Johann's pupil Leonhard Euler. Because there was only one chair of mathematics at the University of Basel, most had to seek employment elsewhere. Johann went to Groningen, Daniel to the Russian Academy of Sciences at St. Petersburg, and Jakob Hermann to Padua, Bologna, Frankfurt-an-der-Oder, and then to the academy at St. Petersburg. Euler went to St. Petersburg, to Frederick the Great's Academy of Sciences at Berlin, and then back to St. Petersburg. Euler became the ablest and most productive mathematician of the eighteenth century (some historians would extend that judgment to include all of time). As an analyst he was unsurpassed. The French mathematicians tended to stay put, but Maupertuis left Paris to become president of the Berlin Academy, and Lagrange, who was born in Italy but was partly French, moved from Turin to Berlin and then to Paris. The development of mathematics and mechanics during the Enlightenment was the work of a few individuals, who, although they retained their national loyalties, could work equally well in France, Switzerland, Holland, Russia, Prussia, or Italy.

This great burst of mathematical analysis came to a close with Lagrange's *Analytical mechanics* (1788). Lagrange carried rational mechanics to the highest point of generality and abstraction that it was to reach during the Enlightenment. The introduction to his book proudly boasts, "There are no figures in this book. The methods that I demonstrate here require neither constructions, nor geometrical or mechanical reasoning, but only algebraic operations, subject to a regular and uniform development."[6]

Try as he might, however, Lagrange could not divorce mechanics entirely from the physical world. His mathematical analysis was necessarily built on physical concepts, but then as now physical concepts developed much more slowly than did new mathematical techniques. Physical concepts are extraordinarily difficult to create. Mathematical methods may not be any simpler, but when they are needed they seem to be found much more quickly. A physical concept is usually defined intuitively in its early stages and means different things to different scientists. During the process of resolving these differences the concept becomes more precise, as does its range of application. Only when it is finally given a quantitative form and a precise means of measurement does its meaning become fixed.

Two such concepts that were studied in rational mechanics during the Enlightenment were *vis viva* ["living force"] and "action." *Vis viva* was thought by its creator, Leibniz, to be the dynamic quantity that was conserved in the universe. It guaranteed that the universe would never run down. The eighteenth century witnessed numerous arguments as to whether *vis viva* corresponded to any real or meaningful thing and whether it was in fact the quantity that was conserved in the universe.

Whereas *vis viva* was a measure of God's desire to conserve his creation, "action" was a measure of his efficiency. God, as a perfect being, was expected to act always by the most economical means, and therefore the "action" in any motion in the universe should be a minimum. The "principle of least action" stated just that – that in any motion the action consumed (measured by the product of the mass, the velocity, and the distance) would be a minimum. The concepts of both *vis viva* and action had philosophical and theological roots in the idea of the economy and simplicity of nature. In the eighteenth century *vis viva* and action took on more specific meanings, and they received their final formulations in the nineteenth century: the conservation of *vis viva* became the conservation of energy in the 1840s, and the principle of least action became the foundation of analytical mechanics, which reached its most general form in the work of William Rowan Hamilton (1805–65) and Carl Gustav Jacob Jacobi (1804–51) in the 1830s.

The concept of *vis viva* had been introduced in 1686 by Leibniz in an article entitled "Brevis demonstratio erroris mirabilis Cartesii" ["A brief demonstration of a notable error of Descartes"]. Descartes had argued in his *Principia philosophiae* [*Principles of philosophy*] (1644) that the "quantity of motion" in the universe must remain the same, but he had defined the quantity of motion as the product of mass and velocity. Descartes's successors, and in particular the great Dutch physicist Huygens, had shown that Descartes's principle of the conservation of motion was valid only if the products of mass and velocity were understood to be directed quantities: In other words, motions in opposite directions were to be subtracted one from the other. With this correction, Descartes's "quantity of motion" is equivalent to our modern principle of the conservation of momentum. But Leibniz argued that although the concept of the conservation of the quantity of motion, understood as a directed quantity, was valid, this was not the quantity that kept the universe from running down. That quantity, according to Leibniz, was the product of the mass and the square of the velocity – not just the product of the mass and the velocity, as Descartes had said. Leibniz called the quantity that he arrived at *vis viva.* Because it was a scalar quantity, and therefore not dependent on the direction of the motion, Leibniz described it as an "absolute" quantity. When this measure of the force of bodies in motion is used, opposite motions do not subtract, and therefore the quantity of force in the universe appears to be preserved.

Throughout the eighteenth century, scientists debated whether quantity of motion or *vis viva* was the proper representation of the force of a body in motion. Voltaire, d'Alembert, and the Dutch physicist Willem Jakob 'sGravesande (1688–1742) all claimed that the controversy was merely a dispute over words, because both principles were valid if used correctly. Their sentiments did not settle the debate, however. At first glance, the conservation of *vis viva* seemed to be valid only for totally elastic collisions, or changes of motion that took place gradually. In inelastic collisions, *vis viva* appeared to be lost, but Leibniz argued on metaphysical grounds that it was not truly lost. He believed that although it was lost to the body as a whole it still remained in the motions of the tiny parts of the body.

Although the controversy seemed to be primarily an intellectual exercise, the most decisive progress came from experiment. Giovanni Poleni (1683–1761) in Italy and 'sGravesande in Holland did experiments in which they measured the impact of a falling ball by dropping it onto clay. When they dropped balls of different masses but of the same size, the same impressions would be made in the clay only if the lighter ball were dropped from a greater height. In fact, they found that in order to obtain equal impressions it was necessary to drop the balls from heights inversely proportional to their masses. From Galileo's law of falling bodies it was known that the heights would be proportional to the squares of the velocities at impact. The experiments showed that if dents in clay were used to measure the force of a body in motion, force would have to be

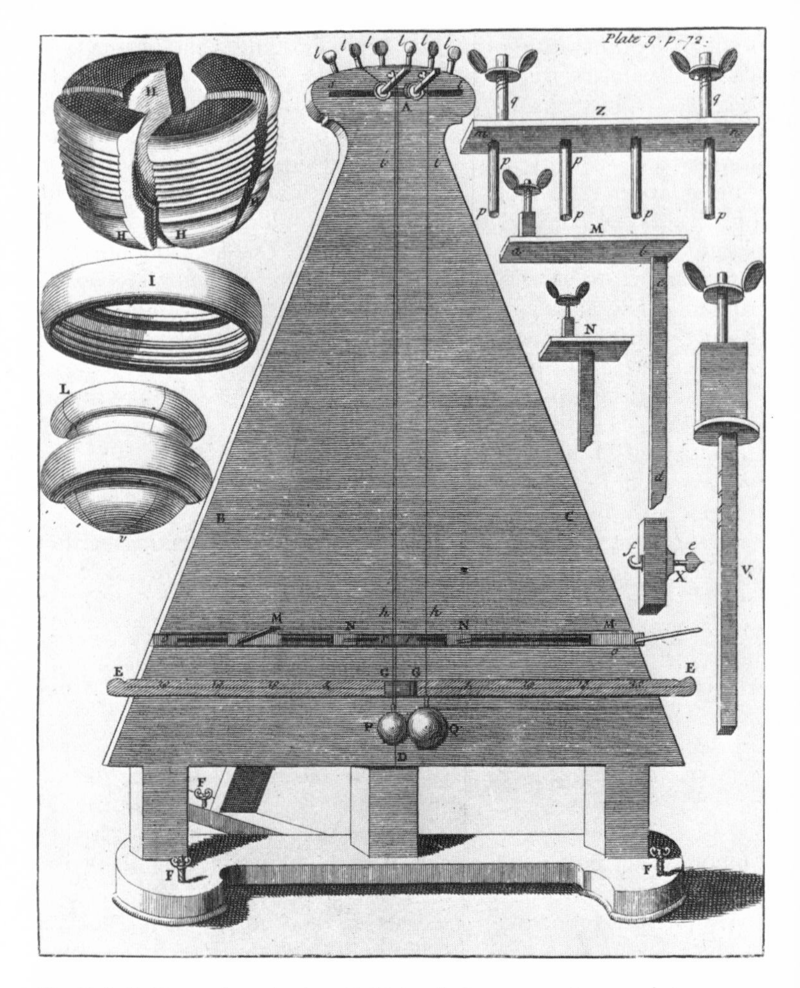

Fig. 2.4. Evidence for *vis viva.* Collision balls mounted as pendulums soon became the standard apparatus for displaying the laws of collision. The apparatus shown in this illustration was employed by Willem 'sGravesande. He built an even more complex instrument on which he could perform oblique as well as direct collisions. In the upper left hand corner is a ball that could be disassembled and loaded with soft clay to demonstrate inelastic collisions. He also built devices containing springs and ratchets that would compress and store the "Force" of the collision in the spring. With such devices 'sGravesande was able to determine which effects were proportional to the quantity of motion (momentum) of the colliding balls and which were proportional to the vis viva (kinetic energy). *Sources:* Willem

understood as *vis viva.* As a result of these experiments, 'sGravesande came out in support of *vis viva* in 1722 and concluded that "what was before only a dispute of words now becomes a dispute about real things" (see Figure 2.4).

The conversion of 'sGravesande confused the ideological debate because he was one of the leading supporters of Newtonian philosophy on the Continent. Newton himself had never attached any significance to *vis viva.* He agreed with Huygens that the product of the mass times the square of the velocity was a number that remained constant in elastic collisions, but for him it had no physical meaning. Moreover, Newton believed that the universe would run down if it were not for God's intervention to renew his creation. It was God who kept the universe going, not any abstract physical principle. According to Leibniz, *vis viva* was the force in a moving body. Since Newton denied that any such force existed and admitted only forces exerted *between* bodies, it is not surprising that his followers joined with the followers of Descartes to deny the importance of *vis viva.*

In the early years of the Enlightenment, the strongest support on the Continent for Newton's philosophy came from Holland. At Leiden, 'sGravesande, Hermann Boerhaave (1668–1738), and Pieter van Musschenbroek (1692–1761) elaborated the scientific method that they had observed in Newton's work and in the process created a new school of experimental physics. In 1720, 'sGravesande published his enormously successful *Physices elementa mathematica, experimentis confirmata, sive introductio ad philosophiam Newtonianam [Mathematical elements of physics confirmed by experiments, or introduction to the philosophy of Newton]* (1720), which expounded Newtonian philosophy with experimental examples. It drew great praise from the English mathematicians, who were understandably dismayed two years later when 'sGravesande came out in support of *vis viva* as a significant concept in mechanics.

Mechanics and Enlightenment Philosophy

At the center of this turmoil over the proper measure of force was an extraordinary woman whose life and career exemplify many

---

Caption to Fig. 2.4 *(cont.)*
Jacob van 'sGravesande, *Mathematical elements of natural philosophy confirmed by experiments, or an introduction to Sir Isaac Newton's philosophy,* 2d ed. (London, 1721), vol. 1, p. 72, pl. 9. By permission of the Syndics of Cambridge University Library.

aspects of the Enlightenment. She was not especially prominent as a natural philosopher nor was she the main protagonist in the *vis viva* controversy, but her writings had great influence, and her life touched those of most of the contributors to the debate over forces. This was Gabrielle de Breteuil, Marquise du Châtelet. She came from a noble family; her father was chief of protocol at the royal court. In 1725 she married the Marquis du Châtelet, and while he served as governor of the city of Semur-en-Auxois she had three children. She then saw little of the marquis when he chose to pursue a military career elsewhere. In 1733 she joined forces with Voltaire – as his mistress, to be sure, but also as his companion in the study of science and literature. Voltaire, of course, became the leading literary figure of the Enlightenment, and in 1734 he published his *Lettres philosophiques* [*Philosophical letters*], the book that launched him on his career.

The *Philosophical letters* was a product of Voltaire's visit to England in the years 1728–9. He had gone there in exile after two sojourns in the Bastille for writing scurrilous verses and for directing his wit against people in power. Disillusioned with his native land, he found things vastly better in England and said so in his letters, which were highly critical of French customs, laws, commerce, and society. The *Philosophical letters* can be thought of as the first bombshell of the French Revolution, because it began the critical attack that was used by the revolutionaries to justify their uprising in 1789. In England Voltaire learned about the work of Newton and consequently devoted a large section of the *Letters* to science. For many Frenchmen the *Philosophical letters* was their introduction to Newton.

After his return from England, Voltaire sought advice from 'sGravesande in Holland and from the leading French Newtonian, Maupertuis, in order to avoid errors in the scientific portions of his book. Maupertuis was the first to openly advocate the Newtonian theory in the Paris Academy of Sciences and the first on the Continent to use Newton's theory of gravitation for determining the shape of the earth, obtaining a result very different from that predicted by the system of Descartes. His *Discours sur la figure des astres* [*Discourse on the shape of the heavenly bodies*] (1732) appeared just at the right time to support the letters on science that Voltaire was revising for his book. Madame du Châtelet became captivated by the new science and began in 1734 to study with Maupertuis and with his younger associate at the academy, Alexis Clairaut.

The publication of the *Philosophical letters,* however, brought the authorities down on Voltaire, who preferred leaving Paris to another stay in the Bastille. He and Madame du Châtelet went to her chateau at Cirey where they could study science and literature without being disturbed.

In addition to Maupertuis and Clairaut, Madame du Châtelet and Voltaire had many other important visitors, including Francesco Algarotti (1712–64). In 1735, Algarotti was in the process of writing an account of Newton's optics called *Il newtonianismo per le dame* [*Newtonianism for the ladies*], which later became one of the most popular of all the many popularizations of Newton's philosophy. Voltaire then set to work writing his own account, which appeared in 1738 as *Elémens de la philosophie de Newton* [*Elements of the philosophy of Newton*]. Madame du Châtelet had helped Voltaire with his *Eléments,* but her commitment to Newton was not unquestioning. She learned about Leibniz in 1736, probably from manuscripts that Voltaire had received from Frederick the Great of Prussia and that contained translations of some of the writings of Leibniz's chief disciple in Germany, Christian Wolff (1679–1754).

Once exposed to Leibniz's thought, Madame du Châtelet began to look favorably on *vis viva,* much to Voltaire's dismay. Reading Johann Bernoulli's "Discours sur les lois de la communication du mouvement" ["Discourse on the laws of the communication of movement"] in 1738 persuaded her even further, especially when Maupertuis agreed that Bernoulli's arguments were essentially correct. And to complete the conversion, her next mathematical assistant in 1739 was Samuel König (1712–57), who had gone with Maupertuis to Basel to study with Johann Bernoulli and had also studied with Wolff. In 1740, Madame du Châtelet published her *Institutions de physique* [*Institutions of physics*], a book that more than any other spread Leibnizian ideas in France. Her conversion to *vis viva,* however, did not mean a complete rejection of Newton. In fact her next project was a French translation of the *Principia,* which she began in 1747 with the aid of Clairaut. It remains the only French translation of the *Principia.*

In the meantime, Maupertuis was being wooed by Frederick to come to Berlin as president of his new academy. Frederick liked all things French and on the advice of Voltaire had invited Maupertuis to be the next president of his academy. Maupertuis visited Berlin in 1740, only to be captured at the Battle of Mollwitz and taken to Vienna as a prisoner of war. (While Frederick was leading his armies against the Austrians, Maupertuis was looking for him and accidentally got on the wrong side of the line.) Finally, in 1745, Maupertuis went to Berlin as president of the Académie Royale de Berlin [Berlin Academy of Sciences], where he inaugurated his presidency with an important paper entitled "Les lois du mouvement et du repos déduites d'un principe de métaphysique" ["The laws of movement and rest deduced from a metaphysical principle"]. The metaphysical principle was his "principle of least action." At Berlin, Maupertuis had to attempt to arbitrate a continuing dispute in the academy over the philosophy of Leibniz and Christian Wolff. Wolff's leading opponent was the mathematician Leonhard Euler, who, along with Maupertuis, saw materialist tendencies in Wolff's writings.

In 1749, Madame du Châtelet, who was finishing her translation and edition of the *Principia,* died in childbirth. Voltaire, grief-stricken by her death, sought solace by visiting Frederick in 1750, and in the same year Samuel König joined the academy. Thus Madame du Châtelet's old friends Voltaire, Maupertuis, and König were again all together, but it turned out not to be a happy reunion. Possibly the problem was jealousies left over from the days at Cirey; possibly it was the great deference shown Voltaire by the king; or possibly it was a continuation of the debate over Leibniz and Wolff's philosophy that König supported and Maupertuis opposed. In any case, in 1751 König attacked Maupertuis's principle of least action, claiming to have proof that the idea originally came from Leibniz. The academy was thrown into a turmoil. Frederick sided with his president, and Voltaire sided with König. Maupertuis, in his *Essai de cosmologie* [*Essay on cosmology*] (1750) had claimed on metaphysical grounds that the principle of least action proved the existence of God. Voltaire attacked Maupertuis with his supreme polemical skill in a vicious satire entitled *Diatribe du Docteur Akakia* [*Diatribe of Doctor Akakia*] (1752). Maupertuis fell ill and left Berlin to recover. Voltaire, who found the atmosphere at the royal palace suddenly very chilly, left for the French-Swiss border where he carried on his campaign for enlightenment in a more secure environment.

Maupertuis's principle of least action had serious weaknesses as he had stated it (not the least being his tendency to redefine the term depending on his use of it), and his claim that the principle provided a metaphysical proof of the existence of God was not well founded (the action used always takes an extreme value, but sometimes it is a maximum, not a minimum). Still, the theory did not deserve the derision that Voltaire heaped upon it. Cleaned up mathematically by Euler and shorn of its theological trappings, it became the fundamental principle of analytical mechanics. Action and *vis viva* (or "energy," as it later came to be called) were not recognized by Newton as important concepts, yet they have displaced Newton's concept of "force" from the center of modern physics. Their ascendancy was not complete during the Enlightenment, but it began then and made remarkable progress, in spite of the apparent dominance of Newton's ideas.

Three Tests of Universal Gravitation

While the mathematicians of the Enlightenment were attempting to extract mechanical principles from their metaphysical trappings, they also exploited Newton's law of universal gravitation to obtain new quantitative results. Celestial mechanics was the ideal testing ground for the new theories because it represented one place where theoretical predictions could be precisely tested. The physical problems on which rational mechanics was built were mostly "thought" problems dealing with ideal, frictionless cases. Such cases could not be realized on earth, but the heavenly bodies presented a case where ideal motion was real and quantifiable. And because the motions of the heavenly bodies could be measured very precisely, they provided the ultimate test of the laws of motion and the theory of gravitation. D'Alembert explained the advantage of Newton's quantitative theory of gravitation over Descartes's qualitative theory of vortices:

> The [Newtonian] system of gravitation can be regarded as true only after it has been demonstrated by precise calculations that it agrees exactly with the phenomena of nature; otherwise the Newtonian hypothesis does not merit any preference over the [Cartesian] theory of vortices by which the movement of the planets can be very well explained, but in a manner which is so incomplete, so loose, that if the phenomena were completely different, they could very often be explained just as well in the same way, and sometimes even better. The [Newtonian] system of gravitation does not permit any illusion of this sort; a single article or observation which disproves the calculations will bring down the entire edifice and relegate the Newtonian theory to the class of so many others that imagination has created and analysis has destroyed.[7]

During the Enlightenment, Newton's theory of gravitation was subjected to three dramatic tests: It was applied to the problems of calculating the shape of the earth, the motion of the moon, and the return of Halley's comet.

The debate over the shape of the earth began with Maupertuis's *Discours sur la figure des astres* [*Discourse on the shape of the heavenly bodies*] (1732). In the *Principia* Newton had argued that the rotation of the earth about its axis should cause it to bulge at the equator and to be flattened at the poles, and as evidence he cited the pendulum measurements made in 1672 near the equator at Cayenne by Jean Richer (1630–96). Richer had found that pendulums of the same length swung more slowly at the earth's equator than in France. Newton said that this was because points on the equator were further from the center of the earth than points in France and therefore the force of gravitation was weaker and the pendulums would swing more slowly. According to Descartes, the force of gravitation was caused by a vortex of matter swirling about the earth that caused the earth to be flattened at the equator and elongated at the poles.

In 1718 the leading French astronomer Jacques Cassini (1677–1756) published measurements made by himself and his father Jean-Dominique Cassini (1625–1712) confirming Descartes's prediction that the earth was elongated at the poles. The British scientists William Whiston (1667–1752), John Keill (1671–1721), and John Theophilus Desaguliers (1683–1744) all supported Newton, and a controversy along nationalistic lines began to develop. In 1732, however, Maupertuis spoke out for Newton at the Paris Academy of Sciences and was joined by Clairaut. Maupertuis and Clairaut were the best mathematicians at the academy, and their arguments counted for much, in spite of their relative youth. Still, Cassini's measurements supported Descartes, and the only way to prove them wrong was to make more accurate and more direct measurements.

The major purpose of the Cassinis had been to measure the size of the earth. If the earth were a sphere, as they assumed it to be, then the curvature would be the same everywhere and could be measured wherever it was most convenient. Therefore the Cassinis' measurements had all been made in Europe. They indicated that the curvature of the earth changed between southern and northern Europe, but in order to determine the shape of the earth more precisely it would be necessary to measure the curvature at those places where it differed the most – that is, at the equator and at the poles. An expedition to the equator was suggested in 1733 by Charles-Marie de la Condamine (1701–74). It was approved by the French government and departed for Ecuador under La Condamine's direction in 1735. In 1736 a polar expedition under the direction of Maupertuis and Clairaut sailed for the Arctic Circle, accompanied by the astronomer Pierre-Charles Le Monnier (1715–99), the clockmaker Charles-Etienne-Louis Camus (1699–1768), the Swedish astronomer Anders Celsius (1701–44), and a few other investigators possessing special skills and hardy temperaments. Algarotti was invited, but Madame du Châtelet urged him not to go, and he stayed behind to write *Newtonianism for the ladies.*

These two expeditions encountered incredible hardships. Maupertuis's group had hoped to make their measurements on the ice or from islands in the Gulf of Bothnia but were unable to get enough elevation for sighting between the survey points, and they were forced to place their signals on mountains in Lapland. The expedition to the equator placed its signals on the summits of the Andes. Since the scientists of both groups were more familiar with Parisian drawing rooms than with inaccessible mountain ranges, their success required extraordinary fortitude and persistence. The polar expedition completed its work relatively quickly, its members returning to Paris in 1737 clad in furs and accompanied by several ladies from Lapland. Voltaire dubbed Maupertuis the great "earth flattener" (this was before the König affair) and praised his vindication of Newton. Ten years elapsed before the equatorial expedition arrived back in Paris. La Condamine reached the Atlantic by crossing the Andes and sailing down the Amazon River. A comparison of the results from both expeditions confirmed Newton's theory. The earth was shaped like an onion, not like a lemon, and the law of gravitation held.

The next test of the law of gravitation came in November 1747, when Alexis Clairaut announced at a public session of the French Academy of Sciences that Newton's law would not account for the observed motions of the moon. The moon had always been a problem. Accurate tables of its motion were of practical importance because they provided a way to determine longitude at sea, but such tables were very difficult to produce. The moon moves much more erratically than the planets; Newton had told John Machin that calculating its motion was the only problem that ever made his head ache, and he understood why: It was because the moon is attracted strongly by two bodies – the earth and the sun, pulling at different angles to one another – whereas the planets are attracted strongly only by the sun. The attraction of the planets for each other is much less significant. Calculating the moon's motion, then, required solving the problem of three bodies mutually attracting each other. As calculus progressed, the competition to get the first solution to this "three-body" problem was fierce. In particular, d'Alembert and Clairaut in Paris and Euler at the Berlin Academy all developed methods of approximating the solution.

It is possible to write the equations for three mutually attracting bodies with differential calculus, but it is impossible to solve them directly. The solution can only be reached by successive approximation, and d'Alembert, Clairaut, and Euler had their own favorite methods. All three, however, came to the same startling conclusion – that the observed motion from month to month of the lunar apogee (the point on the moon's orbit farthest from the earth) differed from the predicted value by a factor of two. Without this unanimity it is doubtful that Clairaut would have dared claim that Newton's law of gravitation was in error. The fact that Euler came up with the same result bolstered Clairaut's confidence enough to encourage him to make his dramatic pronouncement.

The response was vigorous, especially since Clairaut had made the trip to Lapland only ten years before in order to confirm the law of gravitation. The great naturalist, Comte de Buffon, claimed on metaphysical grounds that the law of gravitation could contain only one term and therefore had to be as Newton had written it. The controversy continued until May 17, 1749, when Clairaut again, in an equally dramatic announcement, claimed that Newton was right after all. The mathematicians had all made a common error. It was a mathematical simplification that looked at first sight as if it would have had little effect on the result, but on closer inspection Clairaut found that it was responsible for the large discrepancy in the calculations of the motion of the lunar apogee.

D'Alembert quickly agreed with Clairaut, although Clairaut had published only his results without giving his method. Euler was desperate to find the solution and persuaded the Russian Academy of Sciences to announce that the prize for 1752 would be awarded to the best paper on the movement of the moon's apogee. This finally smoked out an answer from Clairaut.

On November 14, 1758, Clairaut made yet another dramatic pronouncement at a public session of the French Academy. He predicted the return of Halley's comet, claiming a margin of error of only one month. Edmond Halley (ca. 1656–1743) had said that the comet of 1682 that now bears his name would return late in 1758 or early in 1759. The precise date of its return was difficult to predict because the path of the comet would be strongly affected by the attraction of any large planet near which it passed. Clairaut, with the aid of Joseph Lalande (1732–1807) and Nicole Lepauté, the wife of a famous clockmaker, set out to calculate the date of return more precisely, using the new theories of perturbation. When, in 1759, the comet reached its perihelion right on schedule (or at least within the thirty-day margin of error), Clairaut was hailed in the public press as a new Newton, much to the chagrin of his rival d'Alembert, who claimed that Clairaut's feat had been "an operation more tedious than difficult" and no evidence at all of the superiority of Clairaut's methods over his own.[8]

These three famous predictions were evidence of the increased refinement of celestial mechanics and of the growing public interest in science during the Enlightenment. The theories continued to be developed in the second half of the eighteenth century by Lagrange and Pierre-Simon Laplace (1749–1827). Laplace finally settled one vexing problem that had disturbed Newton. This was the question of the stability of the solar system. In addition to the cyclical variations in the motions of the moon and the planets, there were also "secular" variations in their motions that accumulated through time. If these variations were to accumulate long enough, the planetary system might become unstable and fly apart. In five important papers written between 1785 and 1788, Laplace demonstrated that the apparent secular variations in the motions of the planets were actually long-term periodic inequalities that could be accounted for by gravitation and that there were thus no "runaway" parameters. The solar system was stable. In his enormous *Mécanique céleste* [*Celestial mechanics*] (5 vols., 1799–1825), Laplace put the final capstone on the mathematical astronomy of the Enlightenment by summarizing and extending the work of his predecessors.

Positional Astronomy

These great analytic feats depended on the more mundane work of the positional astronomers, whose observations became increasingly accurate during the eighteenth century. Tycho Brahe's (1546–1601) famous observations of Mars made without the telescope had been accurate to about 2 minutes of arc. By 1725, graduated arcs fitted with telescopic sights and filar micrometers were accurate to within 8 seconds of arc, and by the end of the century they were accurate to within 1 second of arc (120 times more accurate than Brahe's measurements; the diameter of the moon is about 30 minutes of arc, or 1,800 seconds of arc). Likewise, the equation that had to be applied to all observations to correct for the refraction caused by the earth's atmosphere improved over the century from an uncertainty of as much as 10 seconds of arc to less than 1 second. These technical improvements meant that observations could refine theory and vice versa.

The real leaders of positional astronomy were James Bradley (1693–1762), Johann Tobias Mayer (1723–62), Nicolas-Louis de Lacaille (1713–62), and Jean-Baptiste-Joseph Delambre (1749–1822). In 1728, Bradley explained a puzzlingly erratic movement of the so-called "fixed" stars that sometimes reached as much as 20 seconds of arc. The shift corresponded to the motion of the earth about the sun, and Bradley correctly concluded that it was the result of the fact that a rapidly moving telescope (and any telescope on the earth is moving rapidly) must be inclined slightly in the direction of its motion in order for light from the stars to pass through it. Because the speed of light is finite, it takes time for it to pass through the telescope, and the light does not emerge in the center of the field of a telescope that is directed exactly at a star. The correction was very slight but was large enough to be detected by the improved instruments of the eighteenth century.

In 1748, Bradley found another periodic fluctuation in stellar observations of about 7 seconds of arc. This fluctuation was not the same as the precession of the equinoxes, which was known to be caused by the fact that the earth's axis precesses slowly, like the axis of a spinning top. This new fluctuation was a small wobble on top of the precession. Bradley correctly concluded that it was due to nutation. Like the axis of a spinning top, the earth's axis not only precesses in a smooth, circular motion but also bobs up and down. Since all astronomical observations are made from the moving earth, Bradley's aberration and nutation were needed to correct all accurate astronomical measurements. In 1749, d'Alembert published his *Recherches sur la précession des équinoxes et sur la nutation de la terre* [*Researches on the precession of the equinoxes and on the nutation of the earth*], which solved the mathematical problem for both precession and nutation, although not in a very clear or straightforward manner. As usual, Euler straightened out d'Alembert's tortuous mathematics, gave a much more elegant statement of the solution, and extended the theory to the motion of rigid bodies in general in his famous *Theoria motus corporum solidorum seu rigidorum* [*Theory of the motion of solid and rigid bodies*] (1760).

The very best lunar tables at mid-century were made by Tobias Mayer, who came to astronomy through cartography and therefore concentrated on the practical problem of describing the moon's motion without worrying about mathematical purity. Using Euler's theory and a large amount of observational data, he produced tables that made it possible to determine longitude at sea to within approximately 1 degree. In 1765, three years after Mayer's death, the British Board of Longitude awarded two thousand pounds, a portion of the Longitude Prize, to his widow in recognition of his contribution. The same search for precision characterized the work of Lacaille (who made the first extensive astronomical observations of the southern skies from the Cape of Good Hope) and Delambre, whose statistical methods provided the basis for calculating tables throughout the nineteenth century. By 1800, theory and observation had converged to bring astronomy to a new level of accuracy.

Physical Astronomy

Calculation and observation were the two necessary ingredients for the improvement of astronomical tables, but toward the end of the century telescopes with greater light-gathering power made it possible also to discover something about the physical structure of the universe and the astronomical objects of which it was composed. This was largely the work of William Herschel (1738–1822), originally a musician in the band of the Hanoverian Guards, who came to England in 1757 and was able to find sufficient musical employment (as a church organist in Bath) to sustain his passion for astronomy. Herschel designed and built the largest reflecting telescopes of his era, culminating in a 40-foot-long reflector carrying a mirror 4 feet, 10 inches in diameter (Figure 2.5a). In 1781, using a smaller telescope, he observed a new "nebulous star" that appeared to have a substantial disk. Its motion among the stars soon revealed that it was in fact a planet, the first to be discovered in historic times. The discovery of Uranus brought Herschel a fellowship, a Copley Medal from the Royal Society, and a substantial salary from King George III, on the condition that he devote himself entirely to astronomy.

Working with his sister Caroline (1750–1848), who became a prominent astronomer in her own right, Herschel set out to determine the shape of the universe by mapping the density of distribution of the stars in different directions (Figure 2.5b). For this purpose he had to assume that the stars had approximately the same intrinsic brightness. The difference in apparent magnitude of the stars he then took to be only a function of their distance from the earth. However, with his larger telescopes he soon discovered exceptions, such as stars of differing brightness that were obviously rotating around each other. In addition to these double stars he observed star clusters and nebulas composed of what appeared to be uncondensed clouds of gas. He was also able to collect enough light from the stars to observe their spectra, although he could not detect the absorption lines. Thus by the end of the century physical and stellar astronomy began to move beyond the traditional astronomy that had concerned itself only with positions and motions, primarily of objects in our solar system.

Celestial mechanics and rational mechanics were suitable adjuncts to the mechanical view of nature that characterized the Enlightenment, but Diderot's warning was justified. As examples of rational philosophical thought, mathematics and mechanics had

Fig. 2.5. William Herschel's forty-foot reflecting telescope and his map of the galaxy. Herschel's largest telescope depicted here was too cumbersome for practical use, and the giant mirror suffered distortion from its own weight. Herschel carried out the observations for his star maps using a smaller telescope. The star map shown is a map of our galaxy, with the solar system located near the center. *Sources:* William Herschel, "Account of some observations tending to investigate the construction of the heavens," *Philosophical Transactions,* 74 (1784), 437–51, and "Description of a forty-foot reflecting telescope," *Philosophical Transactions,* vol. 85, pt. 2 (1795), p. 408. Courtesy of the University of Washington Libraries.

proved their worth, and in astronomy they had brought about major advances; but they had contributed little to the design of real machines or to the improvement of experiment. Diderot was wrong, however, in saying that such improvement would never happen. In the years roughly from 1780 to 1840, the methods and instruments of the experimenter became increasingly quantitative. Mathematics and mechanics found a place precisely where Diderot thought they had no place. Furthermore, the decline of the mechanical philosophy, which Diderot had predicted, did not produce the decline in rational mechanics that he thought would necessarily follow. The analytic methods of the eighteenth century were picked up by the mathematical physicists of the next century and used in electrodynamics, thermodynamics, energetics (the science of conditions and laws governing energy), and a host of nonmechanical fields of physics and chemistry. Nor was he correct in his gloomy assumption that the end of analysis would be the end of mathematics. Every department of mathematics in today's universities proves him wrong.

CHAPTER III

Experimental Physics

By the end of the Enlightenment, experimental physics had come to mean the use of a quantitative, experimental method to discover the laws governing the inorganic world. The original meaning of the term *physics,* however, had been quite different; and as a result the word continued to be used ambiguously throughout the eighteenth century. The discipline of physics had originally been created by Aristotle, and it had nothing to do with experiment or quantitative measure, nor was it limited to the inorganic world. Aristotle's *Physics* treated form, substance, cause, accident, place, time, necessity, and motion through a priori arguments that could then be used to explain the phenomena of the world, both organic and inorganic. In fact Aristotle was more successful in his description of the animal world (also part of physics) than he was in his writings on cosmology or terrestrial motion.

Experiment was almost unknown in antiquity. An experimental tradition did begin in Western Europe during the Renaissance, but it was called "natural magic," not physics. There was also a tradition of applied mathematics, but it was not physics either. It was called "mixed mathematics." During the seventeenth century, physics, as part of speculative philosophy, continued to be taught in the schools in Latin, whereas mathematics, a practical subject with mostly military applications, was taught in the vernacular. Descartes, for example, graduated from college with the impression that mathematics was useful only in the mechanical arts.

Jean d'Alembert argued in the "Preliminary discourse" to the *Encyclopédie* that mathematics was basic to all of physics:

> The use of mathematical knowledge is no less considerable in the examination of the terrestrial bodies that surround us [than it is in astronomy].

All the properties we observe in these bodies have relationships among themselves that are more or less accessible to us. The knowledge or the discovery of these relationships is almost always the only object that we are permitted to attain, and consequently the only one that we ought to propose for ourselves.[1]

Thus, for the mathematically inclined, experimental physics had value only to the extent that its laws could be made quantitative. The importance of mathematics for experimental physics was debated throughout the century. Diderot, the Comte de Buffon, and even Benjamin Franklin (1706–90) condemned the excessive use of mathematics in physics, claiming that it led the scientist away from nature and into a false reliance on abstract forms. The Marquis de Condorcet sided with d'Alembert and claimed that except for the mathematicians, nobody at the French Academy of Sciences was doing any useful work. It was all worthless "physicaille" – parlor tricks and experimental busywork that led nowhere. The debates were about the proper balance between experiment and mathematics. Both were essential for arriving at knowledge, and both were regarded as being in the province of reason. As Voltaire argued in his *Philosophical letters,* the English philosophers Bacon (1561–1626), Locke, and Newton had shown convincingly that knowledge about the physical world could not be obtained from first principles without resort to experiment. Reason dictated a middle course, combining experiment with quantitative measure to allow a constant check on theory.

There had, of course, been famous experimenters in the seventeenth century: Edmé Mariotte (d. 1648) in France, Robert Boyle in England, and Newton, who had combined experiment and mathematics in an especially impressive way in his *Opticks* of 1704. Although Newton was admired both by Continental scientists and by his own countrymen, it was on the Continent, especially in Holland, that natural philosophers elaborated on the experimental Newtonian method. With the installation of William of Orange on the throne of England in 1688, intellectual contact between England and Holland increased. Beginning in 1715, Hermann Boerhaave, Willem 'sGravesande, and Pieter van Musschenbroek, all at the University of Leiden, followed Newton's lead in organizing experiments. Boerhaave, who became a famous doctor and chemist, initiated the Dutch program in his oration of 1715 entitled "De comparando certo in physicis." In that same year 'sGravesande, who was originally trained as a lawyer, went to England as secretary to the Dutch ambassador and met Newton and other British scientists. In 1717 he became professor of mathematics and astronomy

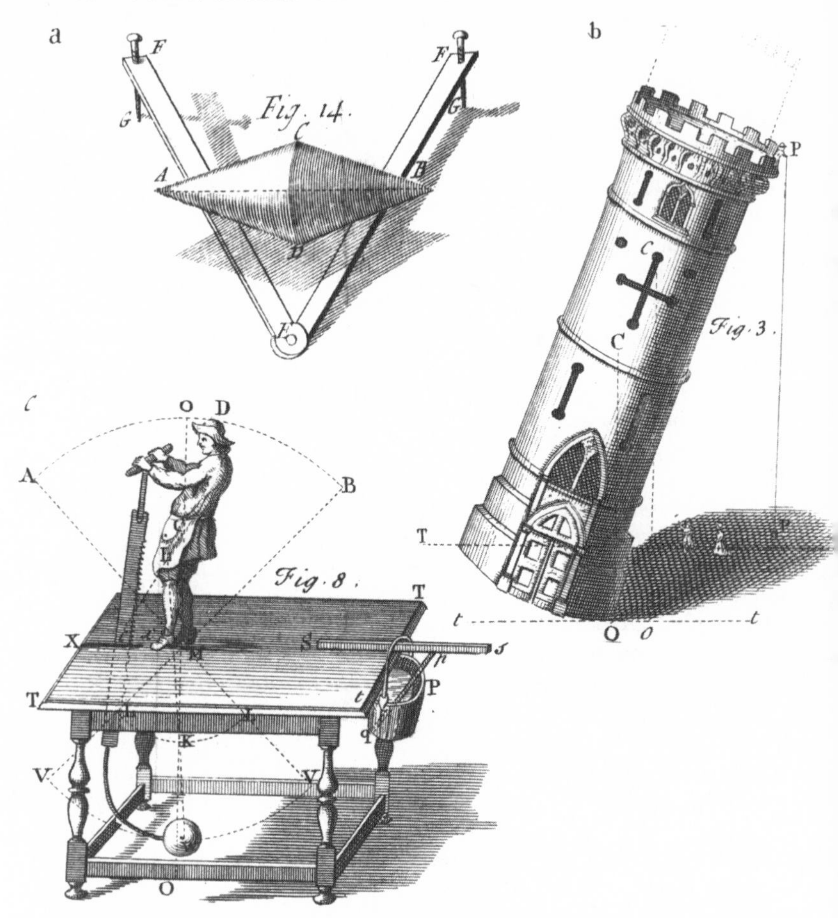

Fig. 3.1. Apparatus for physics demonstrations. A favorite subject for physics demonstration was centers of gravity. The double cone (a) appears to run uphill, but as it rolls toward the separated ends of the tracks its center of gravity is actually being lowered. The model tower (b) will fall over if its center of gravity is not over its base. And a hidden weight helps the sawyer (c) saw his plank. These pieces of demonstration apparatus are still used to instruct and entertain physics students. *Sources:* John Theophilus Desaguliers, *A course of experimental philosophy* (London, 1744), vol. I, pl. IV, fig. 14, and pl. V, fig. 3. By permission of the Syndics of Cambridge University Library.

at the University of Leiden and in 1720 he published his *Physices elementa mathematica experimentis confirmata sive, introductio ad philosophiam Newtonianam* [*Mathematical elements of physics confirmed by experiments, or introduction to the philosophy of Newton*]. It was immediately translated twice by two competing English Newtonians, John Desaguliers and John Keill, and continued through two Latin editions and four more English editions during 'sGravesande's lifetime. 'sGravesande emphasized the importance of mathematics in experimental physics, but it played only a minor role in his work. The emphasis was on experiments and demonstration apparatus, all of which he described in great detail. It is always a surprise to look through the copper engravings at the ends of textbooks by 'sGravesande, Desaguliers, Musschenbroek, Nollet, and other experimenters of the Enlightenment and find demonstration apparatus that with only minor design changes is still used in introductory physics classes (Figure 3.1). The earliest demonstration apparatus set a style that persisted for centuries.

The work of 'sGravesande and the other Dutch physicists redefined physics by making it experimental and by narrowing it to what we now recognize as physical science. From 1720 on, experimental physics commonly included the study of heat, light, electricity, and magnetism but excluded anatomy, medicine, natural history, and chemistry. The French translation of 'sGravesande's book was titled *Eléments de physique* [*Elements of physics*] (1747), Musschenbroek wrote an *Essai de physique* [*Essay on physics*] (1737), and many other books with the word *physics* in the title appeared during the Enlightenment. All of them excluded or greatly diminished the role of the life sciences and emphasized demonstration experiments.

In Germany the situation was slightly different. Experiment thrived there as well, but it was based on the philosophical foundation of Leibniz as altered and interpreted by Christian Wolff. Wolff's *Allerhand nützliche Versuche, dadurch zu genauer Erkentniss der Natur und Kunst der Weg gebahnet wird* [*Generally useful researches for attaining to a more exact knowledge of nature and the arts*] (1721–3) was published the same year as 'sGravesande's *Mathematical elements of physics*. Like 'sGravesande's work, Wolff's book described demonstration experiments and gave detailed instructions for making and using the apparatus, but unlike the Dutch physicists Wolff attempted to create a single rational systematic philosophy, after the model of Leibniz. Wolff's philosophy made little headway in England. In France, however, Madame du Châtelet accepted it in her *Institutions of physics*, and the *Encyclopédie* made use of it in the articles on experimental physics.

The new apparatus and experiments did not immediately make experimental physics quantitative, however, because they were designed to create, not measure, phenomena. Electrical apparatus, for instance, allowed one to discover which objects could be electrified and under what conditions, but it did not measure anything. Measurement had to wait until qualitative theory had specified what it was important to measure. Thus efforts to measure electrical effects came only after experimenters had produced a wider range of new electrical phenomena and had attempted some qualitative theoretical explanations. When, toward the end of the eighteenth century, precise measurement became an important goal of experimental physics, then the imagined "subtle fluids" used to account for phenomena began to be replaced by quantitative laws that made physical phenomena more predictable, if not more understandable.

The Subtle Fluids

The concept of a subtle fluid was a necessary step in the process of quantification. A "subtle" or "imponderable" fluid was a substance that possessed physical properties but was not like ordinary matter. The movement of the subtle fluid carried the physical property with it but did not convey any mass. The best examples of subtle fluids were electricity and heat. Experimenters could easily observe the transfer of electrical effects from one object to another, but they could not detect any accompanying change in weight. Heat, likewise, flowed from a hot object to a cold one without any apparent change in mass (although some experimenters believed that they could detect such a change). It seemed easiest to explain the transfer of electricity or heat by attributing it to a weightless fluid that carried the observed properties. We still use the same image when we speak of heat and electricity "flowing" or assign a heat or electrical "capacity" to an object. Other physical phenomena that could be explained by subtle fluids were gravitation, light, magnetism, and the principle of combustion, although not all of these phenomena lent themselves equally well to explanation in terms of subtle fluids.

The subtle fluids had the added advantage of showing what should be measured in physics. They provided a theoretical framework around which one could build physical concepts like "charge," "electrical tension," "heat," "heat capacity," and "temperature." Newton had hoped to explain all of these physical phenomena by the action of forces between the atoms of matter, in the same way that he had explained the motions of the heavenly bodies by forces of gravitation acting between them. The forces of gravity could be measured, however, whereas there was no way to directly measure the forces acting between the atoms. Therefore a physics based on attractions and repulsions between atoms could only lead to descriptive theories in experimental physics.

The opposite was true of ether or fluid theories. Electricity and heat were found to be conserved, at least in certain situations. Therefore it was convenient to think of them as substances that carried physical properties with them. The density of the fluid was proportional to the intensity of the effect. For example, the thermometer measured the concentration of heat fluid in an object. If heat were not a substance, it was not clear what the thermometer was measuring. Until the concept of energy was developed, the mechanical theory of heat had no simple answer to that problem. All attempts to weigh the fluids of electricity and magnetism failed or were inconclusive, so it was assumed that the subtle fluids carried physical properties but were not themselves material. In one sense they harked back to the "sympathies" and "antipathies" of the hermetists or the "principles" of the alchemists, but unlike those ambiguous, unmeasurable qualities, the subtle fluids were the only way that experimental physics could be made quantitative in the first half of the eighteenth century.

The concept of subtle fluids made its appearance around 1740, when demonstration experiments in physics were rapidly gaining in popularity. At the same time a reinterpretation of Newton's work intended to bring his authority to the support of the new theories was under way. Newton had not made it clear whether the forces acting between the planets and between the parts of matter acted at a distance or through some intervening medium called an "ether." Earlier Roger Cotes (1682–1716) had written a preface to the second edition of the *Principia* (1713), supposedly with Newton's blessing, that described gravity as a force acting at a distance without any intervening medium. Natural philosophers universally accepted this interpretation of Newton's theory of gravity until 1740. After 1740, when the study of electricity and heat began to make progress, ether theories reappeared and were ascribed to Newton. In 1744 two famous letters from Newton to Henry Oldenburg (1676) and to Robert Boyle (1679) that described ether hypotheses were published, and three books in English on Newton's concept of ether appeared between 1740 and 1745. Suddenly Newton's gravity meant action by an ether rather than action at a distance. Thus the subtle fluids were supported in physics not only by new theories but also by a reinterpretation of the old ones.

The concept of subtle fluids was also suggested to some extent by the new interest in "air" during the first half of the century. Stephen Hales's *Vegetable staticks* (1727) ended with a long chapter on the analysis of air "fixed" in bodies. In decomposing a variety of substances by heat, Hales (1677–1761) had been able to collect large quantities of "air" from the reactions. Contemporary chemistry recognized only one element in the gaseous state, and that was the element "air," so Hales regarded all of the gases that he had collected as air. Without searching for any differences in their chemical properties, he merely measured their volumes and wondered how so much air could be contained in such a small solid. To compress the "air" collected back into the volume of the solid from which it came would require great force. How could it be contained without exploding the solid in which it was "fixed"? Hales's emphasis on the repulsive or expansive property of air led naturally to an emphasis on the expansive properties of the even more subtle fluids of heat and electricity. Heat not only caused substances to expand; it also caused solids and liquids to give off "air" that expanded without limit. Electrical fluid also repelled other electrical fluid, so that a charge spread over a conductor and leaped to another conductor that contained a lower concentration of the fluid. Thus, following the model of Hales's air, self-repulsion became a common property of the subtle fluids.

The self-repulsive or expansive property of the subtle fluids was also a property of "fire." The Dutch Newtonians always related the subtle fluid of heat to the element of fire, and "fire," like Stephen Hales's "air," was one of the four Aristotelian elements (earth, water, air, and fire). Fire was the most volatile and least substantial of all the elements; therefore it was the chief agent of change, as witnessed by its role in combustion, fermentation, decomposition, and evaporation. Heat, light, and electricity were all forms of fire, according to Boerhaave and Musschenbroek. 'sGravesande was more cautious in his theorizing but agreed substantially with his Dutch colleagues. Fire existed in all bodies. It could not be created or destroyed but could be transferred from one object to another. In some heat phenomena, such as friction and the focusing of the sun's rays by a burning lens, the fire was not actually transferred but was activated in the body from some latent or unfocused state.

The Dutch physicists also identified electricity with fire. Benjamin Franklin at first followed their lead, referring to the spark as fire until 1750. After 1750, when it became apparent that the properties of heat and electricity were very different, he made a careful distinction between them. Madame du Châtelet and Voltaire both wrote prize essays on fire, and at the end of the century Antoine-Laurent Lavoisier incurred the anger of the revolutionary Jean Paul Marat by criticizing Marat's studies of fire. Thus the old elements of fire and air lay at the conceptual foundations of the subtle fluids. It is easy for us to overlook their importance, since we are equipped with twentieth-century hindsight. Our modern physics attaches no particular significance to the Aristotelian elements, but to understand experimental physics and chemistry as the natural philosophers of the Enlightenment understood them, it is necessary to know "fire" and "air."

The theory of subtle fluids allowed physicists to create new physical concepts and to quantify them, at least to a certain extent. But ultimately the theory became more confining than liberating. Just as Newton discovered that he could describe the phenomena of gravitation mathematically without supposing any ether, so did the physicists of the late eighteenth century discover that they could quantify physical concepts such as temperature, specific heat, charge, and capacitance without assigning any specific subtle fluid to them. At that point experimental physics became more phenomenalistic and more quantitative. The theory of subtle fluids ceased to be of much value, although it had been essential in the earlier stages of theory formation and quantification.

Electricity

Of all the subtle fluids conceived of during the Enlightenment, the electrical fire was the one that caused the most excitement and attracted the most researchers. The study of electricity became the model for experimental physics, both in the kinds of experiments performed and in the construction of apparatus. Electricity had several advantages over the other fields of experimental physics. Once research began in earnest, experimenters rapidly discovered new electrical phenomena, making their work rewarding. The experiments were dramatic, especially after 1746 when the discovery of the Leyden jar made it possible to accumulate very large charges. Electrical experiments also investigated attraction. Newton had constructed an entire world system from the single idea of universal gravitation and had concluded that other attractions and repulsions between atoms would probably account for all of the phenomena of chemistry and physics (using the latter term in its modern sense). Electricity seemed to be the phenomenon that was most likely to exhibit these interatomic forces. By 1733 it appeared that all substances could be electrified by friction and that electricity

Fig. 3.2. Stephen Gray's electrified boy. Stephen Gray and Granville Wheler found that electricity could be communicated over considerable distances if proper substances were chosen for the conductors and for their supports. Among those objects that would conduct electricity was a 47-pound child suspended by silk cords. After Gray's experiment, the electrified boy became a standard part of electrical demonstrations. *Sources:* Johann Gabriel Doppelmayr, *Neu-entdeckte Phaenomena von bewunderswürdigen Würkungen der Natur* (Nuremburg, 1774). Courtesy of John L. Heilbron. Also appearing in *Electricity in the seventeenth and eighteenth centuries: a study of early modern physics* (Berkeley, 1979).

was therefore a universal characteristic of matter. The attractive power of amber rubbed by a cloth had been known to the ancients, but it had also been an artifically produced anomaly that seemed to have little value in explaining nature. The association of electricity with lightning and the discovery that all substances could be electrified moved electricity from the periphery to the center of scientific attention.

Electrical attraction, unlike gravitational attraction, could be controlled in an experiment. It could be increased, decreased, transferred, screened, hidden in bodies, made visible as a spark or a coronal discharge, used to ignite inflammable liquids, to stun animals, and to produce a host of other variable phenomena. Gravitation could not be altered in any way and was too weak to be measured between masses that would fit into the laboratory. When it was discovered that electricity both attracted and repelled, it came even closer to fitting the needs of matter theories that required both an attractive force to account for cohesion and a repulsive force to account for the expansion of gases.

Electrical experiments were popular and suitable for the scientific amateur, but it would be a mistake to assume that amateurs dominated the study of electricity. Benjamin Franklin's kite, electric spider, and lightning bells; Stephen Gray's electrified boy dangling from silk cords (Figure 3.2); the ubiquitous "electrified Venus" with her electrical kiss; and the Abbé Jean Antoine Nollet's (1700–70) spectacular electrification of 180 gendarmes for the edification of the king and queen of France might persuade one that electrical science had only recreational value during the Enlightment, but in fact even the most elaborate showmen were using their experiments to test existing theories and to suggest new ones.

The electrical experimenters found places in the universities and the scientific academies of Europe. A small but significant proportion supported themselves by giving public demonstration lectures. The Jesuits, who had been leaders in experimental physics during the seventeenth century, continued to hold a prominent place until their order was suppressed in 1773. These scientists, who made experimental physics their profession, comprised approximately half of the active electrical experimenters. The rest were artisans, professional men, and the independently wealthy who did not depend on their experiments for their income.

Income was important because scientific apparatus was always expensive. In most cases the professor of experimental physics at a university was expected to provide his own equipment, although this changed as the century progressed. Universities occasionally

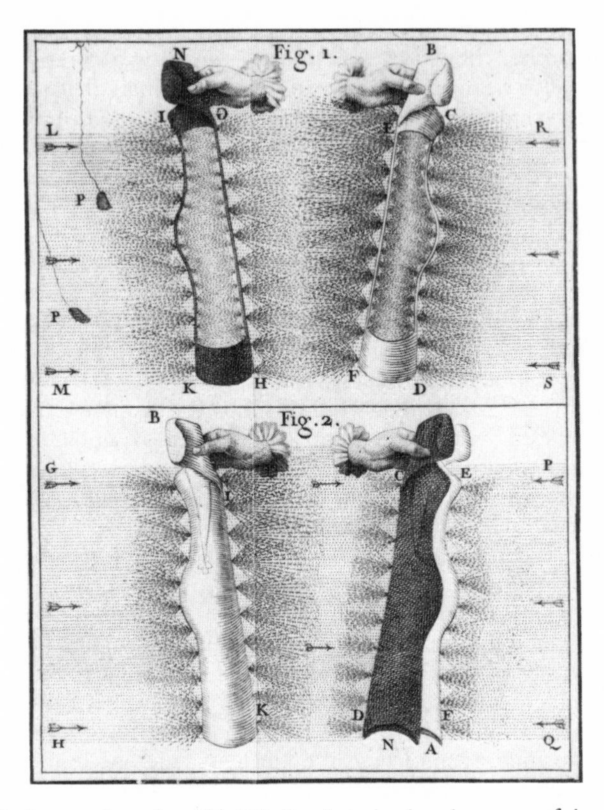

Fig. 3.3. Symmer's socks. Abbé Nollet thought that the cause of the strange behavior of Robert Symmer's silk socks was contrary jets of electric fluid. The jets inside the socks (Figure 1) were supposed to explain why the socks ballooned out when electrified, and the external jets (Figure 2) were supposed to explain why the socks attracted one another. *Sources:* Jean Antoine Nollet, *Lettres sur l'électricité (III) dans lesquelles on trouvera les principaux phénomènes qui ont été découverts depuis 1760* (Paris, 1767). Courtesy of John L. Heilbron. Also appearing in *Electricity in the seventeenth and eighteenth centuries: a study of early modern physics* (Berkeley, 1979).

purchased a professor's apparatus for the use of his successor, but however it was financed, creating and maintaining a substantial *cabinet de physique* was an expensive undertaking. Some of these *cabinets* became very large, the most famous being the collection of the Teyler Foundation in Haarlem. In 1785 the director, Martinus van Marum (1750–1837), ordered a giant electrostatic generator from John Cuthbertson, an English instrument maker, who had moved to Holland. The English still made the best instruments, both for experimental physics and for astronomy. The great electrostatic generator at the Teyler Foundation produced a spark 2 feet long and as thick as a quill pen. Unfortunately van Marum could find few advantages for such a mighty blast in the context of eighteenth-century electrical science. The huge spark did, however, give support to the theory of a single electrical fluid, because it branched like a tree in one direction only, confounding the prevailing French theory that the spark was carried by contrary currents of electrical fluid.

Sometimes a very inexpensive piece of apparatus in the hands of an astute observer could outperform the most elaborate machines. One extraordinary case was "Symmer's socks." In November 1758, Robert Symmer (ca. 1707–63) noticed that when he pulled off his silk socks in the evening "they frequently made a crackling or snapping noise" and emitted "sparks of fire." Experimenting further with his socks, he observed that if he put both a black and a white sock on the same foot and pulled them off together, they exhibited no charge. But if he then pulled the socks apart, they crackled and bulged out as if still occupied by ghostly legs. If the socks were brought back together they collapsed, only to reinflate when separated again. Unlike the spark from van Marum's machine, Symmer's socks suggested the presence of two electrical fluids. If there were only one electrical fluid, the socks should have neutralized each other permanently when they were brought together (see Figure 3.3). Electrical phenomena discovered in increasing profusion and maddening unpredictability bewildered even the most hardy theoreticians.

The Early History of Electricity

William Gilbert (1544–1603) had made the first extensive series of investigations of electricity in his book *De magnete* [*On the magnet*], published in 1600. He was primarily interested in magnetism and investigated electricity only in order to distinguish it from magnetism. In addition to making that important distinction, Gilbert discovered many other "electrics" – substances other than amber that attracted light objects after being rubbed. His "nonelectrics" were substances that he could not electrify by friction. We would now recognize them as conductors, but of course in 1600 Gilbert knew nothing about electrical conduction.

After Gilbert, electrical experiments were carried out by the Jesuits and by members of the Italian Accademia del Cimento. Descartes attempted to include electrical attraction in his scheme of etherial vortices, and Robert Boyle investigated the behavior of electricity in a vacuum. The vacuum provided an important test for electrical theory because it was one way to determine whether air had some mechanical role in electrical attraction or whether an electrical effluvium was entirely responsible for the action. Unfortunately, electrical experiments became more complicated in a vacuum, because gases at low pressure conduct electricity, especially at the high voltages produced by friction. The electrical discharges produced in partially evacuated flasks were quite unlike ordinary sparks and therefore difficult to accommodate in any theory.

It was this glow in a vacuum, the so-called barometric light, that began the train of electrical researches that were so fruitful during the Enlightenment. Francis Hauksbee (ca. 1666–1713), "curator of experiments" at the Royal Society, began research on the luminosity of phosphorus in 1705 under instruction from members of the society. As part of his investigation he studied the phenomenon of barometric light – an occasional flashing that can be observed in a barometer in the vacuum above the mercury. Hauksbee soon discovered that the barometer was not necessary to produce the flashes: Mercury dribbled over a glass surface in a partial vacuum also caused flashes. Very small flashes could even be observed at atmospheric pressure. They grew brighter as the air was pumped out until it was about half gone, then dimmed as the air pressure dropped further. Hauksbee replaced the mercury with other materials rubbed together in a partial vacuum and still obtained the flashes. Finally he discovered that merely rubbing an evacuated globe on the outside was sufficient to produce the flashes. He mounted the glass globe on an axle and caused it to glow brightly when he placed his hands against the spinning globe.

The phenomenon was obviously electrical, although Hauksbee had little chance of explaining what was going on. He was observing a high-voltage electrical discharge through a gas that was under reduced pressure, the principle used today in mercury vapor street lights and neon signs. Although Hauksbee could not explain the barometric light, the discovery of this one unusual and inexplicable effect led to other more fruitful experiments. As we shall see, at the end of the century just such an unusual and complicated phenomenon observed by Luigi Galvani (1737–98) led to the discovery of current electricity.

Hauksbee found that his spinning globe also provided a convenient way of producing electricity. A spinning glass globe or plate became the standard electrostatic generator throughout the century. Hauksbee also rubbed a long tube of glass to produce sparks, and this apparatus became the other standard generator. Hauksbee discovered that a glowing, spinning globe could cause a nearby evacuated globe to glow as well and concluded that the electrical "effluvium" carried about by the spinning globe must be rubbing against the nearby globe, causing it to glow too. He attempted to learn more about this effluvium by hanging threads about the spinning globe (see Figure 3.4). Instead of being blown around by the effluvium as he expected, the threads stood out stiffly, pointing toward the center of the globe.

With a large glass tube he made pieces of leaf brass dance about, first being attracted to the tube, then repelled, then attracted again. Earlier researchers, including Hauksbee's mentor Newton (who was president of the Royal Society during the time when Hauksbee was curator of experiments), had observed this phenomenon but believed that the repulsion was merely a mechanical rebounding of the brass from the tube, not the action of an electrical effluvium. Huygens recognized repulsion, as did 'sGravesande, but Hauksbee was understandably baffled. There did not seem to be any conceivable effluvium, atmosphere, or ether that would explain his apparently contradictory experimental results.

In 1729, Stephen Gray (1666–1736), a dedicated amateur experimenter and occasional contributor to the *Philosophical Transactions* of the Royal Society, discovered that electricity could be communicated over rather long distances by contact. Gray used as a generator a glass tube that he kept corked at both ends to keep out dust. Keeping the dust out was a good idea, because Hauksbee had found that dust in the tube could reduce its power. While he was checking to see if the corks themselves would reduce the tube's effectiveness, Gray was surprised to see the feather that he was trying to attract with the tube move to the cork instead. Since he was preparing to investigate the possible transfer of the "emanation" to other objects, he was alert to the fact that the corks might acquire electricity from the tube. Gray then tried communicating the electricity further, through a stick stuck into the cork and surmounted with an ivory ball. When this succeeded, he fastened to

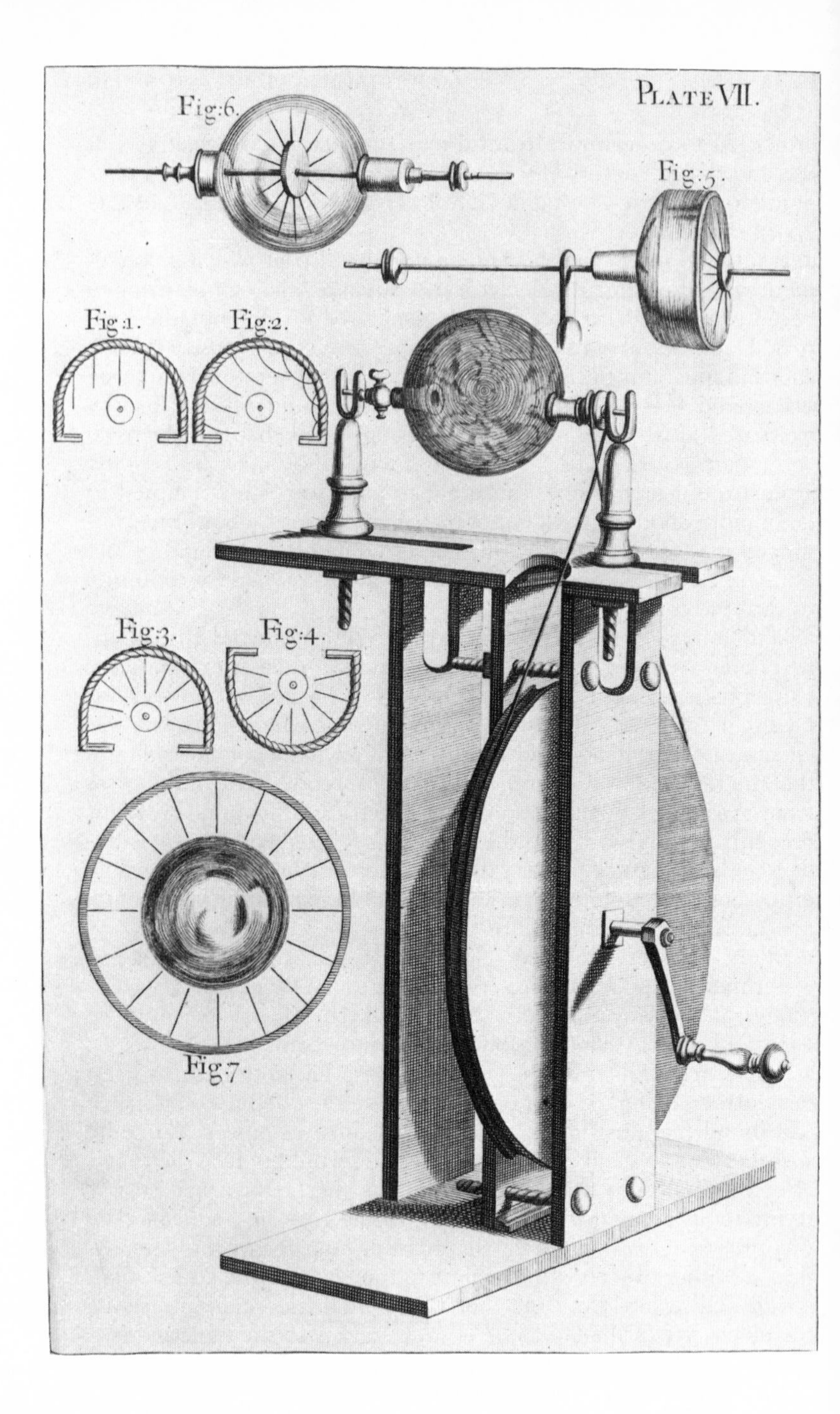
PLATE VII.
Fig:6.
Fig:5.
Fig:1.
Fig:2.
Fig:3.
Fig:4.
Fig:7.

the tube a fishing rod with a string attached, making a total distance of 52 feet, and again he was able to attract objects with the end of the string. Continuing his experiments a month later with a neighboring amateur scientist, Granville Wheler (d. 1770), he managed to carry the electricity 650 feet along a heavy string suspended from silk cords mounted on poles in his orchard.

The discovery of electrical conduction gave evidence of an electrical fluid and presented an opportunity for spectacular demonstrations. Gray and Wheler electrified by conduction a small boy suspended from the ceiling who could then attract objects with all parts of his body.

The next major advance in electrical science was made by Charles-François de Cisternai Dufay (1698–1739), a young infantry officer who pursued his scientific investigations with a command of previous research and a degree of organization that was completely missing in the haphazard experiments of Gray. Expanding on Gray's work in 1733, Dufay systematically experimented with different materials to see which ones could be electrified. He succeeded in electrifying everything that could be rubbed except metals, and these he electrified by induction – that is, by bringing the glass tube close to the object to be electrified (which was placed on an insulated stand), drawing off a charge from the other side of the object, and then removing the glass tube. Dufay discovered that wetting a string made it conduct better, that glass was a better insulator than silk, that attracted bodies were indeed repelled after they struck the glass tube, and, most important of all, that there appeared to be two electricities, not just one. The electricity produced by rubbing a vitreous substance like glass attracted the electricity produced by

---

Fig. 3.4. Francis Hauksbee's investigation of the barometric light. The spinning evacuated glass globe glowed when touched with the hand. In order to detect the motion of the electrical effluvium that he believed surrounded the electrified globe, Hauksbee hung threads inside and outside the glass. But instead of being carried around by the effluvium as he had expected (Figure 2), the threads stiffened and pointed directly at the glass, a result not easily explained by the effluvial theory. After these experiments by Hauksbee, the spinning glass globe became the standard instrument used to create large amounts of electrical charge. *Sources:* Francis Hauksbee, *Physico-mechanical experiments on various subjects. Containing an account of several surprising phenomena touching light and electricity, producible on the attrition of bodies . . . together with the explanations of all the machines . . .* (London, 1709), pl. VII. By permission of the Syndics of Cambridge University Library.

rubbing a resinous substance like amber. Each kind of electricity repelled electricity of its own kind. Dufay called these electricities "resinous" and "vitreous" after the kinds of substances that were rubbed to produce them. Although Dufay never talked about electrical fluids and carefully limited his description of experiments to the phenomena observed, it was natural to assume a two-fluid theory in which each fluid repelled fluid of the same kind but attracted fluid of the other kind. Dufay's collaborator, Abbé Nollet, who became the most prominent French electrician during the Enlightenment, explained the two electricities as opposing currents of the electrical fluid emerging in jets from the electrified body. Nollet's assumption that the electrical effluvium was in rapid motion was natural because the most striking phenomenon of electricity had been from the beginning the way in which small bits of paper or leaf brass were hurled about by the presence of an electrified body. The assumption that a rapidly moving effluvium was responsible for this agitation seemed obvious.

Benjamin Franklin's One-Fluid Theory

An alternative view was that of Benjamin Franklin. Following Newton, whose ideas he had studied in the works of 'sGravesande, Musschenbroek, and Desagulier, Franklin proposed a single static electrical "atmosphere" that attracted and repelled by pressure rather than by the impact of an electrical wind. (Newton's gravitational ether, as he had described it in a letter to Boyle, had been of this type.) Franklin's theory of an electrical atmosphere was never very successful in explaining the phenomena, but Franklin freely admitted that it was nothing but a speculation. His contribution was not dependent on any particular atmosphere, and, like a good disciple of Newton, he made it his object to reduce the phenomena to rule, not to explain them.

In 1743 Franklin saw a demonstration lecture in Boston given by a Dr. Spencer of Edinburgh who repeated Stephen Gray's trick of electrifying a small boy suspended from silk cords. Franklin came away with the impression that he had seen a demonstration revealing "fire diffused through all space." In 1745 the Library Company of Philadelphia received from Peter Collinson (ca. 1693–1768), a Quaker merchant in London, a copy of the *Gentleman's Magazine* containing a lengthy description of spectacular electrical experiments performed in Germany. The article, a translation by the Göttingen physiologist Albrecht von Haller (1708–77) of an earlier article in the Dutch journal *Bibliothèque raisonée,* described experiments at length and clearly identified them as being electrical. Collinson also included a glass tube so that the members of the Library Company could perform their own experiments. Working largely independently from the Europeans, with Franklin as their chief experimenter, the Philadelphians made two important contributions to the growing electrical science. Franklin described them in 1747 in letters to Collinson that circulated in England before they were published in 1751 in the *Philosophical Transactions*.

The first discovery concerned the peculiar power of pointed conductors to "draw off and throw off" the electrical fire. The experimenters had found that if one brought a sharp bodkin up to an electrified metal sphere, the "fire" leaked off the sphere to the sharp point for a distance of up to 8 or 10 inches. In the dark, the experimenters observed a glow at the point, and afterward the sphere was found to have lost its electricity. To the practical-minded Franklin this experiment later suggested the lightning rod, which was intended to defuse the electricity in thunderclouds by drawing it off before it could strike as lightning. It was one of the first inventions to make good Francis Bacon's promise that science would produce new and useful technology. Franklin suggested lightning experiments in 1749 and performed his famous kite experiment in 1752. By that time experimenters in France had already drawn electricity from the clouds and had even performed the kite experiment independently, so it was not original with Franklin. Franklin's kite had a pointed wire attached to the frame of the kite to draw the electricity. The wet kite string conducted the electricity to the key, which was insulated from Franklin's hand by a silk ribbon. From the key Franklin was able to draw off the "fire" and show that it acted just like electricity produced by friction.

This most famous experiment of Franklin's was extremely dangerous. In 1753 the able physicist George Wilhelm Richmann (1711–53), at the Russian Academy of Sciences, was killed when his apparatus drew a lightning bolt from the sky. A debate raged through the rest of the century over the advantages of pointed as opposed to blunt lightning rods. The proponents of blunt rods, led by Benjamin Wilson (1721–88), argued that pointed rods would attract lightning whereas blunt rods, which Wilson believed should be mounted under the roof, would merely carry away a strike but would not attract it. In fact, the shape of the rod makes little difference in its ability to carry off lightning, and Franklin was excessively bold in thinking that his rods defused storm clouds. The height reached by the rod is much more critical than its shape.

A less dramatic but more important set of experiments was one

Fig. 3.5. Testing lightning rods in the great hall of the London Pantheon. In 1772 the Royal Society of London formed a commission to decide whether the Purfleet Arsenal should be protected by pointed or blunt lightning rods. Benjamin Wilson favored blunt rods; Benjamin Franklin, also serving on the commission, favored pointed rods. Franklin won. But in 1777 lightning damaged the arsenal in spite of the pointed rods. King George III supported Wilson in his experiment, shown in the figure, and ordered blunt rods to replace the pointed ones at Buckingham Palace. In the illustration, Wilson's artifical clouds (the large metal cylinders suspended from the ceiling) are menacing a model of the Purfleet Arsenal. *Sources:* Benjamin Wilson, "New experiments and observations on the nature and use of conductors," *Philosophical Transactions* vol. 68, pt. 1 (1778), p. 246. Courtesy of the University of Washington Libraries.

of the first that Franklin performed. In May 1747 he wrote to Collinson, "We had for some time been of opinion, that the electrical fire was not created by friction, but collected, being really an element diffused among . . . and attracted by other matter, particularly by water and metals." Evidence for this theory came from "the impossibility of electrizing one's self (though standing on wax) by rubbing the tube and drawing the fire from it."[2] This experiment, which indicated that the electrical fluid on the glass tube was not created by the rubbing but drawn from the body of the experimenter, suggested four other experiments:

1. A person standing on wax, and rubbing the tube, and another person on wax drawing the fire, they will both of them (provided they do not stand so as to touch one another) appear to be electrized, to a person standing on the floor; that is, he will perceive a spark on approaching each of them with his knuckle.
2. But if the persons on wax touch each other during the exciting of the tube, neither of them will appear to be electrized.
3. If they touch one another after exciting the tube, and drawing the fire as aforesaid, there will be a stronger spark between them than was between either of them and the person on the floor.
4. After such strong spark, neither of them discovers any electricity.[3]

Franklin explained these experiments by assuming that there was a single electrical fluid of which each person in a neutral state had a share. The electrified glass tube acted as a pump in which the fluid could be moved from one body to another. Therefore if person A transferred fluid to person B and both were insulated from the ground, then B would have an excess of the fluid (be charged positively) and A would have a lack (be charged negatively). Franklin's theory of a single electrical fluid that could not be created or destroyed but only transferred from one object to another explained most of the known electrical phenomena. Electrical fluid repelled other electrical fluid but attracted ordinary matter. Electrical attraction, repulsion, sparking, conduction, induction, and Dufay's vitreous and resinous electricities could all be explained by Franklin's theory. (Dufay's two electricities represented on the one hand a lack and on the other an excess of the fluid.)

Franklin missed the fact that negative charges repel, a phenomenon that his theory could not easily explain because it required that ordinary matter bereft of its electrical fluid should repel other ordinary matter, whereas the theory of gravitation required that matter be attractive. Franz Ulrich Theodosius Aepinus (1724–1802) removed this anomaly in 1756 by merely assuming a symmetry in the attractive and repulsive properties of positive and negative electricity, but he could do this only by giving up all the atmospheres and effluvia that had been part of electrical theory before him. By 1756 experimental evidence had already cast doubt on the atmospheres. The study of electricity would have to become more operational. Theory would describe more but explain less.

The Leyden Jar

The most exciting discovery was also the hardest to explain. This was the Leyden jar, devised by the German Ewald Georg von Kleist (ca. 1700–48) but best described by Pieter van Musschenbroek at Leiden, Holland, whence the name of the device. In January 1746, Musschenbroek wrote to René-Antoine Ferchault de Réaumur (1683–1757), his correspondent at the Paris Academy of Sciences: "I would like to tell you about a new but terrible experiment, which I advise you never to try yourself, nor would I, who have experienced it and survived by the grace of God, do it again for all the kingdom of France." Musschenbroek had collected the electricity from a whirling globe in an iron tube suspended from the ceiling by silk. From the end of the tube hung a brass wire that carried the electricity into a flask containing water. Musschenbroek held the flask in his right hand and tried to draw a spark from the iron tube with his left hand. Suddenly his hand was struck with such force that his "whole body quivered just like someone hit by lightning . . . The arm and the entire body are affected so terribly I can't describe it. I thought I was done for."[4]

His courage may have been weakened by the experience but not his curiosity, and Musschenbroek soon discovered that the human hand was not essential; any conductor would do on the outside of the jar, but in order to obtain the huge shock the same person had to touch the outside conductor and the iron tube at the same time. The Leyden jar long remained the most common form of electrical condenser, but other forms, such as "Franklin squares" composed of parallel metal plates, appeared in England soon after. Several Leyden jars connected in parallel could increase the dramatic quality of any lecture demonstration. After he had electrified 180 gendarmes for the entertainment of the king, Nollet shocked 200 Carthusian monks in their monastery. "It is singular to see the multitude of different gestures, and to hear the instantaneous exclamation of those surprised by the shock."[5] Parlor tricks and practical jokes of this type approached lethality.

Franklin pursued an elegant series of experiments designed to probe the mystery of the Leyden jar. He found that just as much

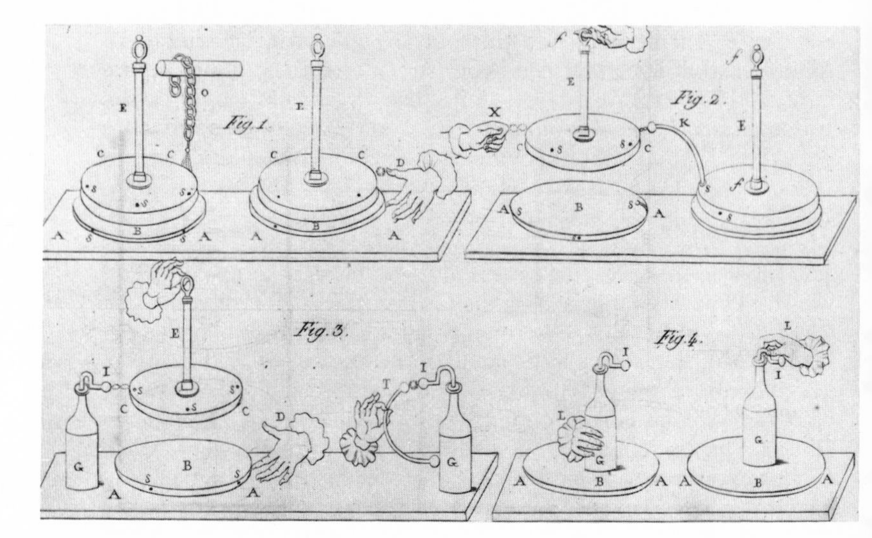

Fig. 3.6. The electrophore of Alessandro Volta. The experimenter could obtain an unlimited amount of electricity from the electrophore merely by lifting and lowering the metal plate (C) with its insulated handle, and by grounding and drawing off the charge at the proper times. With such an apparatus one could charge any number of Leyden jars (G). This version of the separable condenser effectively destroyed the effluvial theories of electricity. *Sources:* Alessandro Volta, *Collezione dell'opere del cavaliere Conte Alessandro Volta* (Florence, 1816), vol. 1, pt. 1, pl. I. By permission of the Syndics of Cambridge University Library.

electricity was pushed off the outside conductor as was taken in by the inside conductor, from which he concluded that the glass of the jar was completely impermeable to electricity. In other experiments he built condensers that could be taken apart in order to locate the residual electricity. He sought to determine whether the charge was on the conductor, in the conductor, or on the surface of the glass. The complexity of the phenomenon and the inadequacy of his theory prevented him from solving the puzzle, but experiments similar to his led to improvements in the theory.

The electrical properties of glass created great confusion for all the early experimenters. Their theories assumed that the electrical fluid existed not only on the electrified body but also in an atmosphere around it. Franklin used smoke and Nollet used fine powder to detect the presence of this atmosphere and to reveal its extent. Electrical attraction and repulsion were assumed to be caused by the direct action of this electrical atmosphere. Glass did not conduct electricity (except when heated), but it did transmit the electrical influence. Hauksbee and Gray could easily attract light objects through several layers of glass. Did the electrical effluvium go through the glass or not? A metal or even a damp cloth conducted electricity, but it screened the effect of attraction. Hauksbee was dismayed to discover that the electrical attraction that could penetrate the wall of a glass bottle could be blocked by a thin sheet of muslin. As long as electrical theorists did not make a distinction between the electrical fluid and its attractive and repulsive influence, they could not explain this particular anomaly. Atmospheres, whether static or in motion, could not be the electricity and its attractive influence at the same time.

The invention of condensers that could be taken apart soon revealed the inadequacies of the concept of atmospheres. Aepinus's air condenser (1756), Johann Carl Wilcke's dissectible condenser (1762), and Allesandro Volta's electrophore (1775) were all disassembled condensers used to locate the electricity and its effect. The electrophore attracted the most attention because it appeared to provide an inexhaustible supply of electricity. Volta (1745–1827) described his *elettroforo perpetuo* to Joseph Priestley (1733–1804) in 1775. It consisted of an insulating cake made of resin and wax set in a metal dish; a metal plate with an insulated handle was placed on top of the cake (see Figure 3.6). The experimenter first electrified the cake by rubbing it or by charging it from a Leyden jar. He then placed the metal plate on the cake and touched the top of the plate to draw off the charge induced by the presence of the charged cake. He then removed the plate by holding the insulated handle. The plate was found to be charged. That charge could be transferred to a Leyden jar and the plate replaced on the cake. The process could be repeated as many times as desired without diminishing the charge on the cake – an apparently perpetual source of electricity. From this experiment Volta concluded that "nothing real" could be passing from the cake to the metal plate; otherwise the electricity on the cake would soon be exhausted. The electricity stayed on the cake, and only a force reached the plate. No atmosphere or effluvium would be inexhaustible. Therefore atmospheres could not explain the electrophore.

With the elimination of the atmospheres, electricians gave up trying to create mechanisms to explain electrical phenomena. Instead they tried to subject these phenomena to quantitative rule. Aepinus's *Tentamen theoriae electricitatis et magnetismi* [*Examination of a theory of electricity and magnetism*] (1759) contained the first successful quantitative analysis of the condenser, but it was regarded with suspicion and little read. Attempts to measure electricity were frustrated not so much by the lack of instruments as by confusion over what should be measured. To separate the concepts of charge, force, tension, and capacitance required a theory that included these concepts. The electroscope, consisting of two threads or two pieces of gold leaf hung side by side, measured charge. The greater the separation of the threads, the greater the charge. (The gunnery metaphor was not accidental. Early electricians believed that they "charged" and "fired" their Leyden jars just as one charged and fired a cannon).

In 1747 Nollet projected the shadow of the threads on a screen containing a protractor scale that allowed him to measure the charge accurately without disturbing the experiment. In 1788 Volta suggested that the charge on a Leyden jar was probably proportional to both the "tension" (intensity of the electricity) and the "capacity" of the jar. Volta had isolated the concepts that would be the most valuable in the quantitative study of electricity. Unfortunately he could not confirm this relationship because of the nonlinearity of electroscopes.

Attempts to determine the force law were more successful. In 1769, John Robison (1739–1805), a student of Joseph Black's at Glasgow, measured the repulsion between charges with an apparatus that balanced the electrical repulsion against gravitational attraction. He was able to show that electrical forces fall off in proportion to the square of the distance between the charged objects, just as gravitational forces do. Charles Augustin Coulomb (1736–1806), a French military engineer, made his much more famous measurements of electrical attraction and repulsion in 1785 by balancing the electrical force against the torsion in a fine wire. The precision of his apparatus set a new standard in experimental physics. After Coulomb, apparatus tended to be carefully engineered for their tasks, not just put together from whatever happened to be lying around the laboratory.

The most successful quantifier of all was Henry Cavendish (1731–1810), but because he resisted publishing his results and wrote only for the fully informed, his accomplishments were not well known. He not only measured electrical force in 1771 but also added the first careful mathematical analysis of experimental error. Cavendish developed a technique for measuring the relative capacitance and the relative resistivity of conductors. The latter he measured by equalizing the jolts he felt upon placing himself in parallel with different lengths of conductors. Through this technique he obtained consistent results, with an error of less than 10 percent.

The Discovery of Current Electricity

At the end of the century the study of electricity was turned in a new direction by a discovery as complex as Hauksbee's barometric light at the beginning of the century. In 1791, Luigi Galvani, professor of anatomy at the University of Bologna, was dissecting a frog in his laboratory. He noticed that when the blade of his scalpel touched the crural nerve in the frog's leg, the leg kicked. It also kicked in unison with an electrostatic machine that was sparking in the room. The kick occurred only when he was touching the blade of the scalpel; no kick occurred when he held the scalpel by its bone handle.

Like the barometric light, the cause of the kick was beyond the scope of any existing theory. Galvani thought he had discovered a new kind of animal electricity and proceeded to check his theory by hanging frogs' legs from brass hooks on an iron trellis outdoors, the purpose being to attract "atmospheric" electricity to the legs. The legs jumped when he pressed the hooks against the trellis, but not because of the atmosphere. As Galvani soon discovered, the contact between the brass hooks and the iron trellis was making the electricity. A single metal produced no kicks.

Galvani must have been extremely observant to recognize these anomalies, but he lacked any adequate theory to explain them. As a physician, he sought a physiological explanation in the anatomy of the frog. His compatriot Volta knew little about frog anatomy but much about electricity. He soon found that the frog served only as a detector of electricity, much more sensitive than any electroscope of the time. Volta soon concluded that the electricity was produced by a circuit containing two dissimilar metals and at least one moist conductor. The electricity from a single metallic junction was weak, but he hoped to multiply the effect by linking several junctions in series. No combination seemed to work until he finally hit upon the idea of stacking up disks of silver and zinc (the metal pair that produced the most electricity), separating each pair with moist cardboard, thereby creating the following series: silver, zinc, cardboard; silver, zinc, cardboard; and so forth, ending in zinc. Just as in the case of the Leyden jar, connecting the top and bottom of this pile produced electricity, but instead of the single spark given off by the Leyden jar, Volta's pile generated a *constant current* of electricity. The pile caused a sensation when it was announced in 1800. Within a year Anthony Carlisle (1768–1840) and William Nicholson (1753–1815) at the Royal Society had used the current to break down water into oxygen and hydrogen. Current electricity led to the whole field of electrochemistry, and the study of electromagnetism followed as a consequence.

The electricians of the eighteenth century had carried their subject a long way. They had refined the theory, quantified their experiments, and improved their apparatus. By the end of the century the concept of the "subtle fluid" of electricity had changed drastically. The concepts of electrical atmospheres and effluvia were gone; electrical fluid was still imagined to flow in conductors, but the forces that it exerted were no longer understood to be the mechanical action of a material in space. As the phenomena of current electricity continued to multiply in the nineteenth century, even the existence of the electrical fluid would be called into doubt. The nature of the subtle fluid changed as the theories of electricity changed, but it was a gradual process. It would be a mistake to say that the discovery of current electricity created a "new science." Even before Volta built his pile, electricians were attempting to measure resistivity, capacitance, electrical "tension," and other quantities that we usually associate with current electricity, and construction of a quantitative theory of electricity was well under way.

Heat and Temperature

Of all the subtle fluids, heat was the one that was most a part of everyday experience (with the possible exception of light). In Aristotle's scheme of things, heat was a *quality,* like color, smell, roughness, or wetness. Qualities could be more or less intense but could not be measured or expressed in numbers. Only *quantities* such as length, weight, and time had magnitude and could be measured. Medieval scholars had talked about different degrees of heat (they counted eight), and scholars at Oxford and Paris in the fourteenth century speculated about the possibility of reducing qualities like heat to numbers, but it was the thermometer, first constructed in 1592 by Galileo, that finally made possible a quantitative study of heat. Galileo's gas thermometer had no set scale and therefore was not really a measuring instrument. The expanding liquid thermometer soon replaced Galileo's gas thermometer, and by 1641 the grand duke Ferdinand II of Tuscany had constructed an expanding liquid thermometer with the end sealed, which was not affected by changes in barometric pressure or by the evaporation of liquid from the tube.

The thermometer scale was completely arbitrary. Some scales were calibrated at a single temperature, and the degree was an arbitrarily chosen distance on the thermometer stem. Others were calibrated at two set temperatures, and the space in between was divided into some number of degrees. Anders Celsius created the centigrade scale in 1742, choosing the freezing and boiling points of water as fixed and dividing the intermediate temperatures into 100 degrees. But Celsius chose 0 degrees for the boiling point and 100 degrees for the freezing point. According to our usage he had the scale upside down, but there is no particular reason why the numbers should run one way rather than the other.

In addition to this arbitrariness of scale, the thermometer *was* what it measured. The experimenter assumed that something called "heat" was proportional to the expansion of mercury. Now that we have other ways of defining and quantifying heat, we know that that assumption was correct for the range of temperatures commonly measured by liquid-expansion thermometers. If it had been an incorrect assumption, the science of heat would have made little progress in the eighteenth century. The best early evidence for the linearity of the temperature scale was the fact that when water is heated over a constant fire, the temperature goes up uniformly.

Of course the thermometer does not measure heat at all, but only temperature. Joseph Black (1728–99), professor of medicine and chemistry at Glasgow and then at Edinburgh, was the first to point out the difference. If one assumed that heat was a fluid, the thermometer measured the intensity or density of the heat fluid in an object, not the total amount of fluid. A 5-gallon container of water contained five times as much heat as a 1-gallon container of water at the same temperature. The thermometer measured merely the density of the heat fluid.

Bacon, Galileo, Descartes, Boyle, and Newton had claimed that heat was not a substance but merely the motion of parts of bodies. In most cases, however, their theories were ambiguous because they believed that the motion of heat was contained in, or was caused by, "fire particles" or some other extremely active special substance. Thus heat was the result of motion, but it was identified with a special substance.

Boerhaave, in his *Elementa chemiae* [*Elements of chemistry*] of 1732, stated the fluid theory of heat unequivocably. He believed that the thermometer measured the density of the heat fluid and that therefore the amount of heat in any object was proportional to its temperature and its *volume,* the object serving merely as a container for heat. Boerhaave was persuaded that the quantity of heat in an object was proportional to its volume by experiments performed by Daniel Gabriel Fahrenheit (1686–1736). Fahrenheit found that when he mixed three volumes of mercury with two volumes of water, the mixture reached an equilibrium temperature halfway between the two initial temperatures. According to Boerhaave's rule, in order to obtain the same equilibrium temperature, the volumes should have been equal. This discrepancy was unfortunate, but at least it was not as bad as the discrepancy between the results of the experiment and the commonly held rule, which was that the heat contained in a body was proportional to its mass. That rule would have required thirteen times as much water as mercury by volume. In an essay written in 1739, George Martine (1702–41), of St. Andrews University, described an experiment in which he heated equal volumes of mercury and water before a fire and found that the temperature of the mercury rose twice as fast as that of the water, demonstrating again that the amount of heat in an object was not proportional to its volume.

Black concluded from these results that the quantity of heat in an object was not proportional either to the volume or to the mass. He argued that different substances had different affinities for or different capacities for heat. A piece of metal felt hotter than a piece of wood at the same temperature because the metal gave off more heat. Therefore not just the volume or the mass but also the nature of the material determined how much heat it contained at a given temperature. Not knowing how the heat was held in matter, whether by chemical combination with the atoms or by saturation of the pores like a sponge, Black in 1760 simply postulated a different heat "capacity" for each substance. By mixing different substances at different temperatures and observing their equilibrium temperature, he concluded that the amount of heat in any object was proportional to the temperature, the mass, and the heat capacity of the object. His only confirmation that the heat capacity was a definite characteristic of a substance came from the discovery that the capacity varied from substance to substance but remained the same for any one substance, whatever the combination of mass and temperature in the experiment.

In 1781, Johann Carl Wilcke (1732–96) came independently to much the same conclusions in Sweden, but instead of heat "capacity" he called the phenomenon "specific heat," in analogy to "specific gravity." For both Black and Wilcke, the specific heat was a constant of proportionality giving the amount of heat required to raise the temperature of a unit mass of a given substance one degree. For every substance the relationship between heat and temperature was different. Water required the most heat for a given temperature change. Very dense materials, which were believed by some to contain the most heat, were often found to have a rather small capacity for heat, mercury being a good example.

Latent Heat

Black and Wilcke were both led to the problem of specific heat by the discovery that a great deal of heat was required to melt ice, even though its temperature remained at the melting point. The prevailing assumption was that ice melted immediately when it reached the melting point. Wilcke recognized the fallacy of this assumption when he tried to melt the snow in a small courtyard by pouring hot water on it. Much hot water melted little snow, even though the snow was at the melting point.

Black saw the same problem in 1757. If snow melted completely when it reached the melting point, the torrents caused by a spring thaw "would tear up and sweep away every thing, and that so suddenly, that mankind should have great difficulty to escape from their ravages."[6] Black placed in a warm room two flasks of water, one frozen and the other liquid, but both at the melting point. The temperature of the liquid water constantly rose, whereas the temperature of the melting ice and water remained at the melting point until all of the ice was gone. From the temperature differences he calculated that it took as much heat just to melt the ice as it took to raise the temperature of an equal amount of water 140 degrees. He checked his results in a more precise experiment in which he added a weighed piece of ice to a weighed amount of warm water.

From the equilibrium temperature, he obtained a comparable figure of 143 degrees Fahrenheit. Black said that this heat was hidden, or "latent," because it did not affect the thermometer. According to the fluid theory, the latent heat was held in combination with the atoms in such a way that the thermometer could not detect it.

Black also measured the latent heat required to boil water into steam. This measurement required a constant source of heat, which Black at first thought could not be obtained, but after a distiller told him that when his furnace was in good order he could tell to a pint the amount of liquor he would get in an hour, Black decided to attempt the experiment. On a constant fire he compared the rate at which water boiled into steam to the rate at which the temperature of cool water increased over the same fire. He discovered that the amount of heat required to boil away a given quantity of water would raise the temperature of that water 810 degrees Fahrenheit if it had not boiled, a figure approximately 20 percent too low but a good result for the rough experiment that Black had designed.

William Irvine (1743–87), who had been a student of Black's at Glasgow, measured the latent heat of fusion of substances such as beeswax and tin and came up with an ingenious theory suggesting a relationship between latent and specific heats. He argued that the heat capacity of an object measured the total heat that it contained, the object acting as a container for the heat fluid. Ice had a measured heat capacity substantially less than that of water. Therefore when ice melted it changed from a small heat container into a large heat container. In order to have water at the same temperature as the ice, the water, with its larger capacity, had to take on a great deal more heat. This was latent heat. Because the heat of vaporization for water was greater than the heat of fusion, Irvine's theory predicted the heat capacity of steam had to be much greater than that of water. In fact it is less. Irvine's theory survived only because it was difficult to determine the specific heat of steam.

The inadequacies of Irvine's simple theory reveal the difficulties in any theory of latent heat. Somehow the heat fluid had to be held in combination with the atoms of different substances in such a way that only a portion of it was detected by the thermometer. How this heat was held latent in the body was difficult to determine. The most simple assumption was that it was held in chemical combination with the atoms of matter, a theory that had important implications for chemistry.

If heat were an actual material substance rather than an "imponderable," it could be expected to have weight, and throughout the eighteenth century many attempts were made to measure it. Boerhaave found no change in the weight of a mass of iron when it was heated. Comte Buffon, however, found that iron gained weight on heating. John Roebuck (1718–94) found just the opposite in 1775, and John Whitehurst (1713–88) confirmed Roebuck's results the next year. They both found that iron gained weight on cooling. Whitehurst warned, however, that the heat from the hot iron being weighed might have caused air currents or uneven expansion of the arms of the balance that might account for the weight difference.

George Fordyce (1736–1802), working with the knowledge of latent heat, realized that a big difference in heat could be obtained with a small difference in temperature, if one compared water and ice. He weighed a flask of water when liquid and when frozen (near the melting point in both cases) and found that it weighed more when frozen, which agreed with Roebuck's and Whitehurst's results and indicated that heat had "levity" (a property that Aristotle had given to the element of fire). In 1787, Benjamin Thompson, Count Rumford (1753–1814), repeated Fordyce's experiment with a more precise balance and varied the experiment to detect anomalies caused by condensation on the flasks and uneven expansion of the balance arms. His results detected no weight at all for heat, which caused him to conclude that heat was a mode of motion, not a substance.

In a more famous set of experiments performed at the military arsenal in Munich of which he was director, Rumford remarked on the great amount of heat produced in boring cannon, particularly if the boring tool were dull. By enclosing the boring apparatus in a box, he was able to boil water from the heat produced as the horses drove the machine. As long as the horses kept moving, the water kept boiling, which seemed to indicate an endless supply of heat.

This experiment has often been described as crucial, since it showed – from our present perspective – that the mechanical theory of heat was correct and that the fluid theory of heat was false. The fluid theory assumed that heat was conserved, which meant that it was not available in an endless supply. Yet however persuasive Rumford's experiments may appear to us in the twentieth century, history refutes this claim. The fluid theory retained its supporters. It could account for conduction and conservation of heat and for change of state in a simple way. It permitted a quantitative science of heat using the thermometer. Even if scientists held the mechanical theory, they would think in terms of a heat fluid for the sake of convenience.

The end of the fluid theory came when scientists sought a theory that would cover radiant heat. It was not easy for either the fluid

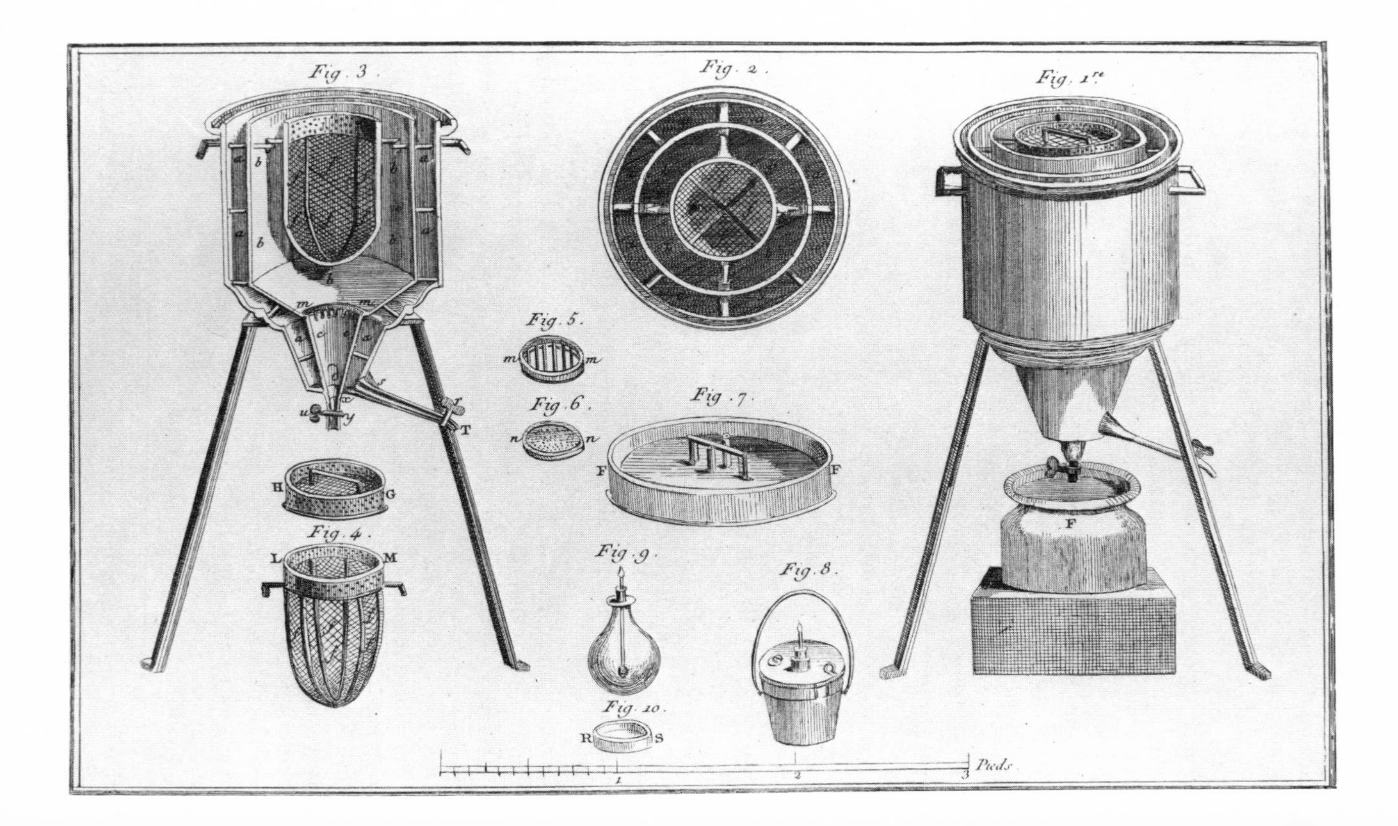
Fig. 3.
Fig. 2.
Fig. 1re
Fig. 5.
Fig. 6.
Fig. 7.
Fig. 4.
Fig. 9.
Fig. 8.
Fig. 10.
F
H
G
L
M
R
S
T
1
2
3
Pieds

Fig. 3.7. The ice calorimeter of Lavoisier and Laplace. This calorimeter used melting ice to measure the heat given off by a body. Ice packed in the outer space of the calorimeter kept the external temperature constantly at freezing. The sample was placed in the wire basket at the center of the calorimeter and the sample's heat melted ice placed in the interior space (*bbbb*). The water from the melting ice ran out of the bottom of the calorimeter and was weighed to determine the amount of heat produced. *Sources:* Antoine Laurent Lavoisier, *Traité élémentaire de chimie* (Paris, 1789), vol. II, pl. VI. By permission of the Syndics of Cambridge University Library.

theory or the mechanical theory to explain the heat from the sun. One could scarcely believe in a fluid flowing all the way from the sun or a mechanical motion acting over that distance. In the nineteenth century the revival of the wave theory of light suggested a comparable wave theory of heat. According to this theory, all heat was radiant. Even heat conduction was merely the radiation of heat waves from one atom to the next. This erroneous theory effectively replaced the fluid theory. It enjoyed a brief reign, to be replaced in turn by the mechanical theory, now fortified by the kinetic theory of gases and the more abstract mathematical formulations of thermodynamics.

The theory of subtle fluids served its purpose well during the Enlightenment. It made possible the quantification of experimental physics and added a more abstract dimension to the prevailing mechanical philosophy. The theory was also amazingly versatile. Franklin's atmospheres and Nollet's effluvia proved inadequate to the task of accounting for all electrical phenomena, but the primary property of the subtle fluid – its conservation – remained. The subtle fluid of heat survived Rumford's experimental refutation because it was simply too valuable to give up. The mechanical theory provided no model for the conservation of heat or its flow from a hot to a cold place. In most cases, giving up the subtle fluid meant giving up the only model that could be understood in simple terms. The subtle fluids were necessary for the quantification of experimental physics in its early years, but as that quantification progressed in the nineteenth century, especially with the creation of new and more precise measuring instruments, the subtle fluids gradually gave way to even more abstract and more mathematical models.

CHAPTER IV

Chemistry

In Herbert Butterfield's famous book *The origins of modern science (1300–1800)* (New York, 1956), there is a chapter entitled "The postponed scientific revolution in chemistry." The title refers to the fact that the new chemistry associated with the oxygen theory of Antoine Lavoisier did not emerge until the 1770s and 1780s, a century after Newton's *Principia* had put the capstone on the Scientific Revolution of the seventeenth century. It is also significant that Lavoisier's contemporaries were conscious of and frequently mentioned a "revolution" that was occurring in chemistry, and that Lavoisier himself stated in 1773, in a private memorandum, that he believed the experiments he was undertaking would "bring about a revolution in physics and chemistry." Scientists at the time and historians since have concurred in identifying chemistry as a subject that enjoyed its "revolution" during the Enlightenment.

In fact, the Chemical Revolution was more the creation of a new science than a change in an existing one. Before 1750, chemistry could not be regarded as an independent discipline. It had long antecedents, but they were ancillary to other fields. Alchemy was a source for many of the recipes and much of the apparatus of chemistry, but this information was concealed in intentionally ambiguous and allegorical language. Alchemy sought to complicate nature, not to rationalize it, and the alchemists' search for the philosopher's stone that would allow them to change base metals into gold was as much a spiritual quest as it was a scientific one. By the time of the Enlightenment, alchemy had all but disappeared.

Chemistry had instead become the business of physicians and pharmacists. The leading chemists – Hermann Boerhaave, Georg Ernst Stahl (ca. 1660–1734), and Joseph Black – were all medical doctors, and most of chemical theory and practice was closely related

Fig. 4.1. A chemical laboratory in the eighteenth century. This chemical laboratory is being used for metallurgy and assaying, which accounts for the many furnaces of different kinds and for the precise balances in the window. *Sources:* William Lewis, *Commercium philosophico-technicum or the philosophical commerce of the arts* (London, 1765), frontispiece. By permission of the Syndics of Cambridge University Library.

to the needs of medicine. As the century progressed, chemistry began to lose some of its dependence on medicine, but the emancipation was slow. Even at the end of the century a chemist would still go to an apothecary to obtain his materials.

Industrial chemistry added to the chemical knowledge that came from medicine. The Industrial Revolution greatly increased the demand for certain chemical products such as alkalis and mineral acids, and the search for improved methods of manufacture resulted in new chemical techniques in metallurgy, ceramics, and textiles, especially in textile dyeing and bleaching. Lavoisier was very much part of that tradition. His research on gypsum, on methods of lighting the streets of Paris, and on gunpowder reveal his practical approach to the study of chemistry.

The mechanical philosophy was another source for the advancement of chemistry in the eighteenth century. Newton had hoped to reduce chemistry to a science describing the mechanical interactions between atoms, but that hope was not fulfilled. Tables of affinity, which were based on displacement reactions and were supposed to show the relative strengths of the attractive forces between atoms, were the most successful parts of that program, but the mechanical philosophy gave no advantage for explaining chemical properties such as acidity, alkalinity, metallicity, salinity, and the chemical operations of combustion, fermentation, and distillation. The Chemical Revolution came about not by any triumph of the mechanical philosophy but by a rationalization of these traditional chemical qualities and operations.

But if the new science of chemistry was not mechanical, it was, nevertheless, physical. The revolution that Lavoisier had predicted in his memorandum of 1773 was supposed to be a revolution in both chemistry and physics. These fields, as we have seen, were not clearly distinguished during the Enlightenment. Lavoisier's colleagues regarded him as being as much *physicien* as *chimiste,* and Lavoisier himself was a leader in the effort to create a position of experimental physics at the Paris Academy of Sciences.

Chemistry and experimental physics were closely associated because they were still based to a large extent on the concept of the Aristotelian elements: earth, water, air, and fire. These four elements were defined more by physical properties than by chemical properties. Earth was solid and held its own shape without a container; water was fluid and took the form of its container; air not only took the form of its container but also expanded to fill it entirely; and fire (in the form of heat) could not be contained at all and passed through the walls of a flask or retort. From our point of view these were physical, not chemical, properties. The most important elements for the Chemical Revolution were air and fire, both of which were believed to be primitive and elemental. Thus the only elastic substance that could be compressed and that would expand to fill its container was air. Air might contain a variety of vapors, odors, and emanations dissolved in it, but the physical properties of elasticity and "expansibility" (the latter a new word coined in the second half of the eighteenth century) came solely from the air, which was the pure, elemental form of atmospheric air.

The Vaporous State of Matter

The crucial realization of the Chemical Revolution was that "air" was not a single element but a physical state that many chemical substances could assume and that atmospheric air was a mixture of several different chemicals in that same "vaporous," "gaseous," or "aeriform" state (again, "vaporous," "gaseous," and "aeriform" were new words invented in the eighteenth century to describe an idea that had only just been recognized). The recognition of the vaporous state meant that chemists for the first time understood that the ability of a substance to completely fill its container did not mean that it was a single chemical element.

Fire was the other important element in the Chemical Revolution. We have already encountered it in the experimental physics of electricity and heat. Chemical reactions always involve loss or gain of heat, but so do many physical reactions. The distinctions that we make today between physical and chemical reactions were not easily made during the Enlightenment. It was obvious that "fire" was responsible for putting substances into the aeriform state, but the operations of fire did not succumb to rational analysis as quickly as did those of air. The creators of the Chemical Revolution never fully understood fire. Lavoisier regarded it as a simple substance and renamed it *caloric*. Joseph Priestley and Henry Cavendish continued to use the term *phlogiston* for the action of fire in combustion. It was only with the rise of thermodynamics in the nineteenth century that chemists were able to fit fire into a satisfactory theoretical framework.

Air and fire had one property in common that was discovered only in the eighteenth century: They could both be "fixed" – that is, hidden in solid and liquid substances. Stephen Hales's *Vegetable staticks* (1727) revealed that enormous quantities of "air" could be released by destructive distillation from many solids and liquids, including hog's blood, amber, oyster shells, beeswax, wheat, tobacco, gallstones, urinary calculi, and a great variety of plant material. The fact that air could be "fixed" in a nonelastic state in solid matter was a startling discovery that attracted much new attention to the study of air. Chemists traditionally had paid little attention to air. Their recipes told them how to make solids and liquids, but what went up the chimney was not part of the recipe.

In 1757, thirty years after Hales described his experiments with fixed air, Joseph Black discovered the phenomenon of latent heat (see Chapter III). Latent heat was fire "fixed" in matter, just as "fixed air" was air fixed in matter. When water boiled, great quantities of fire (heat) were fixed in the steam without producing any change at all in the thermometer reading. The analogy suggested a general theory about the fixation of air and fire. Fire was obviously fixed and released in many chemical reactions: for example, it was released in combustion reactions, and it was fixed in those processes in which it took heat from the furnace. Hales released air from solids but could not fix it in solids. The latter operation would not become possible until chemists isolated different "airs" having differing chemical properties and could employ their art to put these airs into chemical combination with other substances to create a solid product. The problem was a complex one, however, and chemists quickly ran across anomalies that made any simple theory of "fixation" of air and fire untenable.

The most striking anomaly was evaporation. In 1756, William Cullen (1710–90), Joseph Black's teacher, published "An essay on the cold produced by evaporating fluids and of some other means of producing cold." Cullen sought to discover why the temperature reading on a thermometer dropped a few degrees when the instrument was removed from spirit of wine (alcohol). The phenomenon had been observed before but never satisfactorily explained. The secretary of the French Academy of Sciences, Jean-Jacques d'Ortous de Mairan (1678–1771), in his widely read *Dissertation sur la glace* [*Dissertation on ice*] (1749) had described the Chinese refrigerator that cooled water in a porous earthenware jug by evaporation from the jug's outer surface. De Mairan had recognized the cooling effect of evaporation, but only for water. Cullen found that many liquids cooled on evaporation: The more volatile the liquid, the greater the amount of cooling. He also evaporated liquids in a vacuum. Others had noted that water in a vacuum boils at a lower temperature, but Cullen also noted a cooling effect. Placing a pan of ether in a water bath inside the receiver of his vacuum pump, he was able to freeze the water by the cooling effect of the evaporating ether.

These experiments were perplexing, because if liquids evaporated by being dissolved in the air, as was commonly believed, then the removal of the air should make evaporation impossible – but just the opposite occurred. Also, the cooling effect of evaporation indicated that the vapor carried off heat or matter of fire. Possibly the cause of evaporation was not the liquid's dissolving in the air but the liquid's combining with the matter of fire to create a vapor. If this were the case, then it would be fire, not air, that made the evaporated fluid elastic. Might not air itself, then, be a liquid combined with fire, and might not there be many different "airs," all different chemical substances but all in the same aeriform state because they were all combined in the same way with the matter of fire? If this were so, the atmosphere would be a mixture of many "airs," not just the one elemental air of Aristotle. The implications of the experiments on evaporation made by Cullen and others – including Johann Gottschalk Wallerius (1709–85), Johann Theodore Eller von Brockhausen (1689–1760), Abbé Nollet, Charles Le Roy (1726–79), Antoine Baumé (1728–1804), and ultimately Antoine Lavoisier – were not immediately obvious. Yet the anomaly existed, and the old solution theory would not account for it.

The person who recognized these implications most clearly was not a chemist or physicist but a famous French public servant, Anne Robert Jacques Turgot (1721–81), *intendant* for the *généralité* of Limoges and later director general of finances for all of France. Turgot presented his ideas in the article "Expansibilité," which he contributed anonymously to the *Encyclopédie.* The article was unusual in that the word *expansibilité* did not exist in the French language. Turgot coined it in order to name a property of air. Robert Boyle had referred to the "springiness" of the air in 1660, but this word and its French equivalent *ressort* implied that the air was somehow composed of little springs. The expansibility of air was not like that of a spring. Air expanded without limit, whereas springs stopped expanding when they were fully extended. Turgot concluded from the experiments of the Abbé Nollet and Wallerius on evaporation in a vacuum that expansibility was a property not only of air but of all substances in a vaporous state. This "vaporization" (another neologism of Turgot's) would happen to almost all substances if the temperature could be pushed high enough. The heat attached itself somehow to the parts of matter and overcame the attractive forces, causing the parts to separate from one another without limit. "Thus, water, for example, could be made to pass through all three states, solid, liquid, and vaporous, by changing the temperature."[1] According to Turgot, the expansibility of vapor was caused by the fire attached to its parts, and this was so even for air. Hales's air, when fixed in matter, was air from which the fire had been removed and had therefore lost its expansibility. It remained "fixed" in solid or liquid matter until fire was again adjoined to it to make it expansible.

Turgot's theory of expansibility accounted for the boiling of liquids in a vacuum and for the cooling effects observed by Cullen, but it did not explain why evaporation takes place at low temperatures. Even ice will evaporate slowly, without melting, at temperatures below the melting point. If heat joining to the parts of ice causes it to evaporate, why does it not melt the ice first? Turgot returned to the old solution theory to explain this phenomenon. He believed that ice evaporates from the surface and that water below the boiling point also evaporates from the surface, in both cases by being dissolved in the air. Water above the boiling point, however, turns to vapor without the presence of air. Thus boiling occurs throughout the liquid, not just at its surface. Turgot called the transfer to the gaseous state "vaporization" and carefully distinguished it from evaporation, which required the presence of air. Evaporation remained an unsolved puzzle until the advent of kinetic theory in the nineteenth century. Nevertheless, Turgot's limited answer was important for the Chemical Revolution.

Turgot had stated a theory that with considerable elaboration by practicing chemists would lead to the oxygen theory of combustion. A crucial realization was the idea that the different physical states of matter are not to be associated with any specific chemical elements but are all possible for any chemical if it could be carried through a great enough range of temperatures.

The recognition (one might almost say the discovery) of the gaseous state and its relationship to heat seems more like a problem in physics than in chemistry. But acknowledging the existence of the gaseous state was a prerequisite for explaining combustion, the central problem of the Chemical Revolution. Lavoisier began his experiments on combustion in the fall of 1772. Sometime in the summer of that year, before he began his experiments, he wrote a memorandum that he called "Système sur les élémens" ["System of the elements"]. In that memorandum he adopted a position close to Turgot's and used the expression "fluid in vaporization" to describe the vapor produced when water is heated in a vacuum. The use of the new word *vaporization* indicated that Lavoisier had gotten the idea from Turgot's article; in 1775 he referred to the article explicitly.

Gas Chemistry

The argument that chemical substances could exist in a "vaporous" state was reinforced by the discovery of several new "airs." As we have seen, Stephen Hales was the first to collect and systematically measure the air released from solid and liquid substances. He demonstrated that large quantities of "air" could be released from matter by heat, but he did not demonstrate or even consider that the air he collected might be different in different experiments. The first person to identify a new air different from common atmospheric air was Joseph Black, the same Scottish professor who contributed so much to the study of heat.

Black studied chemistry with Cullen at Glasgow, serving for three years as his assistant. In 1752 Black went to study at the University of Edinburgh, and in 1754 he submitted as his dissertation for the M.D. degree a paper entitled *De humore acido a cibis orto et magnesia alba* [*On the acid humor derived from food and on magnesia alba*]. The next year he described his experiments in an expanded form to the Philosophical Society of Edinburgh, which published them in 1756 under the title "Experiments upon magnesia alba, quicklime, and some other alcaline substances."

The medical purpose of Black's research had been to find a medicine that would dissolve kidney stones. The medicine preferred by the Edinburgh professors for this complaint was limewater (calcium hydroxide), made by heating limestone or cockleshells (calcium carbonate) in air to make quicklime (calcium oxide), which, when dissolved in water, became limewater. Two eminent professors, Robert Whytt (1714–66) and Charles Aston, were already studying the properties of quicklime and limewater, disagreeing fiercely over the proper method for producing them and the causes of their causticity.

The production of quicklime from limestone in lime kilns was an old and important industrial process. Quicklime was used in the manufacture of cement and in the production of other alkalis, which in turn had important industrial uses such as the manufacture of soap. So the debate had significance beyond the use of limewater in medicine.

Rather than join in contest against his professors, Black chose to study another mild alkali, magnesia alba ("white" magnesia, or magnesium carbonate), which had only recently been used as a medi-

magnesia alba [$MgCO_3$] ⟶ magnesia usta [MgO] + fixed air [CO_2] ↑

magnesia usta [MgO] + acid [$2HCl$] ⟶ magnesia salt [$MgCl_2$] + water [H_2O]

magnesia alba [$MgCO_3$] + acid [$2HCl$] ⟶ magnesia salt [$MgCl_2$] + fixed air [CO_2] ↑ + water [H_2O]

Fig. 4.2 Joseph Black's experiments on magnesia alba. The chemical names are Black's; the chemical symbols are modern.

cine. Black experimented to see if it had the same properties as chalk, which it closely resembled. Heating the magnesia alba produced a product, magnesia usta (magnesium oxide), that was somewhat like quicklime, but it was not caustic or soluble in water like quicklime. More significantly, Black noticed that the magnesia alba lost a great deal of weight when it was heated, and he commenced a series of careful quantitative experiments to determine what substance had been lost in the reaction and where it had gone. The magnesia usta formed by heating the magnesia alba did not effervesce when acid was added to it. However, the original magnesia alba effervesced vigorously when treated with acid. The effervescence appeared to be air released from its fixed state in the magnesia alba. In both cases the addition of acid to either the magnesia alba or the magnesia usta produced the same magnesia salt (see Figure 4.2).

Black discovered that exactly the same weight was lost when the magnesia alba was heated and then treated with acid as was lost when the magnesia alba was treated with acid directly. Recalling Hales's experiments, he concluded that the loss of weight in both cases was caused by air that had been fixed in the magnesia alba leaving the reaction. In one case it was driven off by heat, in the other case by the action of acid. He then carried out comparable experiments on chalk and showed that a large quantity of air was also released in the roasting of limestone.

At first Black did not collect the air released from his experiments but found its weight from the difference between the weights of the reactants and the products in his experiments. The purpose of the experiments had been to discover the cause of alkalinity, and Black concluded that all of the mild alkalis (carbonates) like magnesia alba contained fixed air, whereas the caustic alkalis (hydroxides) did not. The alkalinity was therefore not caused by the absorption of fire particles from the furnace, as Whytt had supposed, but was inherent in the magnesia and lime, the fixed air acting to reduce the alkalinity.

Black's studies of causticity had shown that air changed the chemical properties of substances in which it was fixed and changed them in very specific ways. Moreover, he remarked in a letter to Cullen written early in 1754 that the air he had discovered had unusual properties of its own. It had a characteristic odor (when he produced it by adding acid to chalk), and it extinguished a flame as effectively as if the burning material had been dipped in water. In further experiments he found that the fixed air was denser than ordinary air and would turn limewater milky. Using the limewater test, he showed that this fixed air would not support life and was itself a product of respiration. It was also given off in alcoholic fermentation and in the combustion of charcoal. Here was an "air" with very special chemical properties. It was not common air but "one particular species only." Black continued to use Hales's name of "fixed air" rather than invent a new one, but as a result of his experiments "fixed air" became the name for a particular air (carbon dioxide), not for atmospheric air or for any other "air" that might be released from a chemical reaction.

Once alerted to the possibility that there might be different airs with different chemical properties, British chemists proceeded to find them. Cavendish isolated "inflammable air" (hydrogen) in 1766, finding that it burned explosively and was much less dense than common air. He also checked the solubility of inflammable and fixed airs and was the very first scientist to store fixed air over mercury instead of water, because of carbon dioxide's high solubility in water.

Joseph Priestley, the most successful searcher for "airs," collected his airs over mercury in an improved version of the pneumatic trough (Figure 4.3). Beginning with the study of fixed air, he obtained the gas in a variety of different ways and showed that its solubility in water increased with pressure. In June 1772 he published a pamphlet entitled "Directions for impregnating water with fixed air" that gave instructions for making artificial seltzer water, previously available only from mineral springs. The supposed medical benefits of this new water (it was believed to prevent scurvy) drew the attention of French chemists, who were ignorant of Black's and Priestley's previous achievements in pneumatic chemistry. In March of the same year, Priestley had begun to present to the Royal Society the results of his pneumatic experiments, which were published late in 1772 as "Observations on different kinds of air." He collected "nitrous air" (nitric oxide), "marine acid air" (hydrogen chloride), and later found "alkaline air" (ammonia), "vitriolic acid air" (sulfur dioxide), "phlogisticated nitrous air" (nitrous oxide, or "laughing gas"), and "dephlogisticated air" (oxygen). Priestley's names

Fig. 4.3. Joseph Priestley's apparatus for studying airs. Priestley's *Experiments and observations on different kinds of air* (1774) contains this drawing of his apparatus. Notice that Priestley not only collected airs in his cylinders but also checked their effects on plants and animals. The inverted drinking glass and the cylinder on the perforated plate confine mice, and the cylinder on the far right contains a plant. *Sources:* Joseph Priestley, *Experiments and observations on different kinds of air and other branches of natural philosophy, connected with the subject. In three volumes; being the former six volumes abridged and methodized, with many additions* (Birmingham, 1790), vol. I, frontispiece. By permission of the Syndics of Cambridge University Library.

for these last two airs were based on the prevailing phlogiston theory of combustion.

The years 1772 and 1774, in which Priestley discovered these airs, were also the years in which Lavoisier took the most important steps towards creating an opposing theory of combustion. The Chemical Revolution had three main ingredients: The first was the theory of the gaseous or vaporous state; the second was the rise of pneumatic chemistry and the discovery of "airs" with different chemical properties; and the third was the discovery of the nature of combustion.

The Problem of Combustion

In 1660 Robert Boyle published an account of experiments that he had performed with his vacuum pump. The vacuum pump had obvious advantages for investigating the role of the air in combustion. He quickly discovered that removing the air extinguished a flame and killed small animals placed in the receiver of his pump. The same thing occurred, of course, but more slowly, with animals and candles kept in a closed container. These experiments indicated to Boyle that some vital substance in the air was necessary for sustaining flame and life. He also discovered that calcination – the heating of a metal to produce its calx (or oxide) – required the presence of the atmosphere and that the calx weighed more than the metal from which it was produced. All of these experiments indicated that combustion, respiration, and calcination were similar processes and that they used up the air or some vital principle in the air. Boyle's efforts to measure this loss of air, however, were unsuccessful. He attempted to measure the loss of pressure in a container of air containing a mouse. The mouse died, but the pressure remained the same.

In 1674, John Mayow (1641–79) performed the same experiment in a container over water and did measure a loss of air when a mouse breathed it or a candle burned in it. (In Mayow's experiments the carbon dioxide produced by combustion and respiration dissolved in the water, whereas in Boyle's experiments it merely replaced the oxygen consumed.) However there was one substance, gunpowder, that burned without air. Robert Hooke (1635–1702) in his *Micrographia* of 1665 and Mayow in his *Tractatus quinque* [*Five treatises*] of 1674 both concluded that it was the saltpeter, or niter, in gunpowder that provided the vital principle necessary for combustion. Sulphur and charcoal, the other two ingredients of gunpowder, burned in air but not by themselves in a vacuum. Therefore there must be some common substance in the niter and in the air that supported the combustion. Mayow called it "nitro-aerial spirit" or "nitro-aerial particles." Hooke held a very similar theory. He claimed that fire was the "dissolution of the burning body by the most universal menstruum of all sulphurous bodies, namely the Air."[2] It was an old idea coming from alchemy and traceable to the late Middle Ages, but with the experimental researches of Boyle, Hooke, and Mayow it took on greater scientific precision and respectability.

The Phlogiston Theory

In Germany, where chemistry had been closely associated with mining and smelting, there arose a competing theory that gave a more obvious explanation of combustion. Johann Joachim Becher (1635–82), in his *Physica subterranea* [*Subterranean physics*] (1669), identified different kinds of earth. According to his theory, combustible bodies contained "oily earth" that was released during combustion, leaving behind "stony or vitreous earth." Georg Stahl greatly extended Becher's theory in his *Specimen beccherianum* of 1703. He renamed the oily earth "phlogiston." According to the phlogiston theory, the principle of combustion is in the fuel rather than in the air. When charcoal is consumed by fire, the phlogiston is given off, leaving behind only a few ashes. Charcoal must therefore be very rich in phlogiston. Respiration expels phlogiston, as does calcination. A calx can be returned to its metallic state by heating it with charcoal because, as we have seen, charcoal is a rich source of phlogiston, which is absorbed by the calx to produce the metal.

When Stahl asked foundrymen what function the charcoal played in reducing the calx to its metal, he was told that the metal buried itself in the charcoal to escape the heat of the fire. To observe the action of charcoal he heated litharge, the calx of lead, in a furnace, then dropped small pieces of charcoal onto the calx. Wherever the charcoal fell on the litharge, a small bit of lead appeared. It seemed obvious that the charcoal added some ingredient that was absorbed by the calx to create the metal. Therefore metals all contained phlogiston, which explained their common properties of density, luster, and malleability.

Combustion, calcination, and respiration all cease when the air in which they take place is saturated and poisoned by the phlogiston. The air is essential. It acts as a sponge to absorb the phlogiston, but once it is vitiated by phlogiston it loses its elasticity and will no longer support combustion. Unfortunately Boyle had shown that the calx (which contained no phlogiston) weighed more than the metal (which did contain phlogiston) from which it came. Could phlogiston weigh less than nothing? Aristotle's element of fire was supposed to have the principle of levity, and Cavendish's "inflammable air" (hydrogen), which he thought from the way it burned might be pure phlogiston, was less dense than common air, but the concept of negative weight was not an attractive notion in the eighteenth century. Still, phlogiston gave a good qualitative account of the phenomena and was a valuable theoretical tool.

Stahl was one of the leading chemists of his age, and his authority was great, but it would be a mistake to assume that the phlogiston theory had ruled chemistry with unquestioned authority for a long time. The most popular textbooks in France during the first half of the eighteenth century were those of Boerhaave and Nicolas Lemery (1645–1715). Neither book mentioned phlogiston. But around 1750 the increasing demands of industry, especially in metallurgy, brought greater attention to the German chemical texts, many of which were translated into French between 1750 and 1760. In 1756 an edited version of Lemery's *Cours de chimie* [*Course of chemistry*] (1675) appeared with a phlogiston commentary. In 1757, J. F. Demachy (1728–1803) translated Johann Juncker's (1679–1759) *Conspectus chemiae theoretico-practicae* [*Conspectus of chemistry*] (1730) and published it as *Elements of chemistry according to the principles of Becher and Stahl.* Baron d'Holbach, who was responsible for the articles on mineralogy and metallurgy in the *Encyclopédie,* was an advocate of the phlogiston theory, and in 1766 he translated Stahl's *Zufällige Gedancken . . . über den Streit von den sogenannten Sulphure* [*Occasional thoughts on the debate over the so-called sulfur*], probably the clearest statement of the theory. The most important German chemist was Andreas Sigismund Marggraf (1709–82), whose papers in the *Mémoires* of the Berlin Academy during the 1740s and 1750s earned the admiration of the French chemists. The influx of German texts coincided with the revival of French chemistry under Guillaume-François Rouelle (1703–70), who began his famous chemical lectures at the Jardin de Roi in 1742. Most of the leading French chemists studied with Rouelle, who drew heavily on the German chemists and adopted a modified form of the phlogiston theory. When Lavoisier began his experiments on combustion in 1772 (Lavoisier also attended Rouelle's lectures), the phlogiston theory had enjoyed a general popularity in France and England for approximately twenty years.

Lavoisier's Experiments on Combustion

There was very little in Lavoisier's activities prior to 1772 that revealed any interest in combustion. He had studied at the Collège des Quatre Nations, which gave the best scientific education of all the Parisian schools, but had received his baccalaureate degree in law in accordance with his parents' wishes. He continued to follow his scientific interests, however, and studied astronomy with Abbé Nicolas-Louis de Lacaille (1713–62), his former teacher at the Collége des Quatre Nations, and accompanied his botany teacher, Bernard de Jussieu (1699–1777) on the latter's famous botanical tours of the Paris region. Since 1763, Lavoisier had helped the geologist Jean-Etienne Guettard (1715–86), a family friend, with his geologic atlas, and in 1765 he submitted a paper to the Paris Academy of Sciences on the mineral gypsum and its use in making plaster. In the same year he also entered the competition for a prize offered by the academy on street lighting and won a gold medal. Admitted to the academy in 1768, he soon became involved in another practical question, that of improving the water supply of Paris. Ever since his geologic studies with Guettard, he had been interested in mineral springs, and the debates at the academy over tests for purity of water led him to investigate the ancient idea that when water evaporates a small part of it is converted into earth. Lavoisier heated water gently for 101 days in a "pelican," an apparatus for constant distillation in which the water evaporates, then condenses, and finally runs back to be evaporated again. A small amount of "earth" appeared in the water, but by very careful weighing Lavoisier showed that the substance had not come from the water but from the glass pelican, which had lost weight equal to the substance produced. The "earth" had not been produced by the transmutation of water but by the destructive action of water on glass.

The method that Lavoisier used in his analysis of water characterized much of his subsequent research. Using a very meticulous laboratory technique, he tackled practical problems, always with an eye out for their theoretical significance. He was an untiring organizer, researcher, and promoter of his own ideas and schemes. In the Academy of Sciences he served on numerous commissions studying everything from the conditions of prisons and hospitals to mesmerism and the nutritional value of vegetables. He was largely responsible for the superiority of French gunpowder, a superiority that became important during the revolutionary and Napoleonic wars, and he was a leader in the effort to reform the system of weights and measures. He carried out agricultural experiments on his own farm at Fréchines and urged a reform of the tax system, which had made improved agricultural methods too costly to use. He also purchased a part interest in the "tax farm," the group of financiers who obtained from the French government the right to collect taxes. In exchange for this right, the tax farm paid the government a fixed sum that was negotiated each year. Although Lavoisier worked efficiently and did not take an unreasonable profit, his colleagues were not always so scrupulous. Friends of the king had managed to obtain official positions on the farm as sinecures, which meant that without doing any labor they reaped enormous profits from taxes that were extracted from the French peasantry. It was Lavoisier's position as a tax farmer, more than any of his numerous other activities, that created enmity toward him during the French Revolution and ultimately led to his execution.

Lavoisier's early research had required precise experimental technique, but it had not been highly theoretical and had had no direct bearing on the problem of combustion. We know, however, that by the summer of 1772 Lavoisier was thinking about air and was familiar with Turgot's theory of the vaporous state of matter. Two other academic chores brought the problem of combustion to Lavoisier's attention. The first involved the destruction of diamonds. It had long been known that diamond, the hardest mineral known, could be destroyed by heat, but it was not certain whether the diamond burned or evaporated in the furnace. Lavoisier, along with the chemists Pierre-Joseph Macquer (1718–84), Louis-Claude Cadet (1731–99), and Mathurin-Jacques Brisson (1723–1806) performed experiments on diamonds at the highest temperature available. This temperature was reached by the great burning lens of the Academy of Sciences. Their experiments eventually showed that diamonds burn in air but do not evaporate. However, these results were not in hand by the time Lavoisier began his experiments on the combustion of sulfur and phosphorus.

In the meantime a second academic chore had fallen to Lavoisier. That was to review a paper on phosphorus by the pharmacist Pierre-François Mitouard. Mitouard found that phosphorus increased in weight when it burned, and he suggested that this increase might have something to do with the air. Lavoisier, who decided to carry out his own experiments on combustion, went to Mitouard to obtain phosphorus.

Lavoisier had also learned more about the calcination of metals by the time he began his experiments in September 1772. In the spring of that year the academy had published Louis-Bernard Guyton de Morveau's (1737–1816) *Digressions académiques* [*Academic*

According to the Phlogiston Theory

lead ⟶ calx of lead [litharge] + phlogiston

calx of lead + charcoal [a source of phlogiston] ⟶ lead

According to Lavoisier

lead + common air ⟶ calx of lead

calx of lead + charcoal ⟶ lead + fixed air ↑

The Modern Interpretation

$2Pb + O_2 \longrightarrow 2PbO$

$2PbO + C \longrightarrow 2Pb + CO_2$

Fig. 4.4. Descriptions of the reactions that produce the calcination of a metal and the reduction of the calx to recover the original metal, according to the phlogiston theory, Lavoisier's theory, and modern chemistry.

digressions] which contained the first clear demonstration that all metals gain weight on calcination. It had been a disputed point, one that the phlogiston theory could not easily explain, because it meant that the loss of phlogiston produced a gain of weight, not in a few unusual circumstances but in every instance of metallic calcination.

We will never know exactly what combination of ideas and circumstances caused Lavoisier to undertake his research on combustion, but it is clear that the pieces of the puzzle lay all about him. The fragments of knowledge that probably contributed to his inspiration were (1) his awareness of the role of the air in the combustion of diamond; (2) his knowledge of the increase of weight caused by the combustion of phosphorus and the calcination of metals; (3) the availability of some vague and incomplete information about the work of the British pneumatic chemists (probably limited to Hales's experiments on fixed air and Priestley's paper on soda water); (4) Lavoisier's familarity with and Tugot's theory of the vaporous state of matter into which all of the experimental facts would eventually have to be fitted.

On September 10, 1772, Lavoisier wrote in his laboratory notebook that he proposed to investigate whether phosphorus absorbed air when it burned. On October 20 he reported to the academy that this was in fact the case and that the combustion of the phosphorus with the air produced "acid spirit of phosphorus" (phosphoric acid), the absorbed air apparently causing the acidity. He performed the same experiment on sulfur and obtained similar results. These experiments on sulfur and phosphorus were in striking agreement with Guyton de Morveau's experiments on calcination. To see whether air had been absorbed in the calcination of lead, Lavoisier reduced litharge with charcoal and observed that at the moment when the calx changed into metal a quantity of air was released that was at least a thousand times greater than the original volume of litharge employed. Obviously, air was fixed in the metallic calx too (See Figure 4.4).

Lavoisier placed a report containing these observations in the hands of the secretary of the academy on November 1, 1772. It was a *pli cacheté,* or "sealed note," intended to establish his priority without forcing him to publish incomplete results. The sealed note was full of optimism: "This discovery [of air coming from calx of lead] appears to me to be one of the most interesting of those that have been made since Stahl, and since it is difficult not to let escape in conversation with one's friends information that might put them on the right track, I believed that I ought to deposit this note in the hands of the secretary of the academy, to remain sealed until the time when I shall make my experiments known."[3] Lavoisier was beginning to suspect that his discovery that air was absorbed in calcination and combustion could be explained by Turgot's theory of the vaporous state, but identifying the airs involved from their chemical properties would be a major task. In February 1773 he wrote in his laboratory notebook: "The importance of the end in view prompted me to undertake all this work, which seemed to me destined to bring about a revolution in physics and in chemistry."[4]

Once one knows the answer, the solution to the puzzle of combustion and calcination seems obvious. We now know that the phosphorus and lead gain weight because they combine with the oxygen in the air. In the reduction of the calx with charcoal, carbon in the charcoal combines with oxygen in the calx to form carbon dioxide ("fixed air"), which is given off during the reaction, leaving behind the metal. The gas that combines with the metal in calcination (oxygen) is not the same gas that is given off in the reduction of the calx (carbon dioxide). This explanation requires, of course, that there be different kinds of air and that atmospheric air be a mixture of these different kinds. In the 1770s it would have been difficult even to contemplate such a radical idea.

The fact that different gases were absorbed and released in calcination and reduction made it very difficult to discover the role of the air in combustion and calcination. When Lavoisier reduced the calx of lead with charcoal, he was delighted to collect large quantities of "air," but it was fixed air. Fixed air might be supposed to be the air that was originally absorbed from the atmosphere when the metal was calcined, because it was the air given off in the reduction of the calx, but calcination does not take place in an atmosphere of fixed air. Therefore it could scarcely be fixed air that the lead had absorbed.

Oxygen and the Vaporous State of Matter

In attempting to understand the role of air in combustion and calcination, Lavoisier extended Turgot's theory of the vaporous state into chemistry. Because cooling takes place whenever there is vaporization or evaporation, Lavoisier theorized that when a liquid turns into a vapor the liquid must have absorbed fire. He assumed that chemicals in a vaporous state are also in combination with fire and that it is the fire that makes them expansible. To explain the results of his experiments on combustion and calcination, Lavoisier concluded that during these processes the fire that had been combined with the air was released as heat and light and that as a consequence the air lost its expansibility and became fixed in the fuel. Lavoisier did not so much get rid of the phlogiston as move it into the air. According to Lavoisier, it was actually the air or part of the air that burned, because it was from the air that the fire was released.

For some chemists like Macquer it was hard to see what all the shouting was about. What difference did it make to a chemist whether the phlogiston came out of the fuel or out of the air? Probably little, as far as the source of the fire was concerned, but if it meant that combustion was to be understood as a combination of the fuel with air, rather than as a simple decomposition of the fuel, it made a great deal of difference. Because Lavoisier was alert to the probability that a portion of the air combined with metals in calcination, he was the person who was most likely to detect that air and understand its properties. But it was not obvious where he should look for it or how he might separate it from the rest of atmospheric air.

In February 1744, Pierre Bayen (1725–98), master apothecary to the French Army, published an article on the red precipitate of mercury "per se" in *Observations sur la physique, sur l'histoire naturelle, et sur les arts,* better known as Rozier's *Journal* after its editor, Abbé François Rozier (1734–93). Bayen showed that this precipitate was a true calx of mercury (mercuric oxide), in spite of the fact that it yielded mercury even when heated without charcoal. No other metallic calx was known to have this property. Bayen heated the precipitate both with and without charcoal and obtained large quantities of air in both cases. Without testing the airs, he assumed that they were both fixed air, which Lavoisier had already shown to be expelled in the reduction of any metallic calx.

Meanwhile, Lavoisier had continued his experiments on calcination by heating lead and tin in closed retorts. He found that the sealed containers weighed the same before and after heating, indicating that no ponderable "matter of fire" had been carried through the walls of the retorts. When the containers were opened, the air was heard to rush in, and the calx was found to weigh more than the metal. These experiments, which he announced to the academy on November 12, 1774, seemed to him to be conclusive. Air had combined with the metal to produce the calx. Unfortunately the experiments had been done before, much to Lavoisier's chagrin when he found out about it.

On August 1, 1774, Joseph Priestley heated the same calx of mercury that Bayen had investigated in February. The occasion for doing the experiment was Priestley's acquisition of a large burning lens, 12 inches in diameter, with which he proceeded to burn any interesting substance that happened to be lying around his laboratory. He had no theoretical purpose in mind; in fact he later admitted that he could not remember why he had performed the experiment, except for the fact that he had done many such experiments and would have required only the slightest incentive to do this one. To his surprise, when he heated the calx of mercury, it turned into mercury without charcoal being present; he also collected a substantial quantity of air. A candle and a glowing wooden splint placed in this air burned brightly. Priestley thought that possibly he had collected "phlogisticated nitrous air" (nitrous oxide, or "laughing gas"), an air that he had discovered earlier and had found to support combustion.

In October Priestley traveled to Paris with his patron Lord Shelburne (1737–1805) and met several French chemists, including Lavoisier. At dinner he talked about the new air that he had discovered. It is unclear how much Lavoisier could have learned from Priestley. There were doubts about whether the powder that Priestley had heated was a true calx. (Red powders could be obtained by heating mercury in a container open to the air and also by treating mercury with nitric acid. Chemists were not yet certain that these two red powders were the same. During his visit to Paris, Priestley acquired a sample that he knew to be a true calx of mercury because it had been made by heating mercury in the air.) There was also uncertainty about what the new air could be. Lavoisier repeated Priestley's experiment with the burning lens the following month but apparently only to see if heat from the sun decomposed

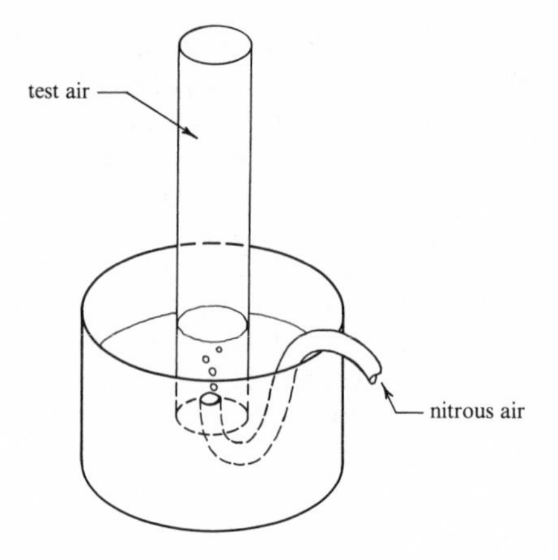

Fig. 4.5. Joseph Priestley's test for the "goodness" of air. The volume of a sample of ordinary "respirable" air collected over water in a glass cylinder decreases by one-fifth when "nitrous air" (nitric oxide) is added to it. The volume does not decrease at all, under the same conditions, if the air has previously been "spoiled" by respiration or combustion. However, the volume of a sample of "dephlogisticated" or "eminently respirable" air (oxygen) decreases by much more than one-fifth when "nitrous air" (nitric oxide) is added to it.

the calx as well as heat from a fire. He did not collect and test the gas. Only in February and March 1775 did he undertake a new series of experiments on calcination, and even then the reduction of the calx of mercury did not figure prominently among his plans.

When Lavoisier finally collected the air from the reduction of mercury calx, he first tested it with limewater and discovered that it was not fixed air, as Bayen had supposed. He then found that it supported combustion and assumed, as had Priestley, that it was phlogisticated nitrous air. But because it supported combustion it might also be common air, and Lavoisier proceeded to apply a test that Priestley had devised for determining the "goodness" of air. The test consisted of adding one volume of nitrous air (nitric oxide) to two volumes of test air over water. Priestley had found that when common air was tested by adding the nitrous air, the volume decreased by one-fifth. (The nitric oxide combined with the oxygen in the test air to produce nitrogen dioxide, which is highly soluble in water.) When air that had been "spoiled" by combustion was tested, there was no contraction at all. (see Figure 4.5).

Lavoisier found that air from the reduction of mercury calx decreased in volume by one-fifth, indicating that it was common air. If he had added more nitrous air than the test called for, he would have discovered that the air from mercury calx contracted even more than common air, but Lavoisier did not add more nitrous air. The test had confirmed his theoretical prediction that the air taken in during calcination of any metal was not the same as the air given off when the calx was heated with charcoal. When he heated the calx of mercury with charcoal, he got fixed air as expected. Fixed air was, then, a combination of common air taken from the calx and carbon from the charcoal. It was a triumph for his theory, because it demonstrated that air was combining with the metal to produce the calx. In his famous Easter memorandum of April 16, 1775, he announced that the air released from the mercury calx was "the air itself, undivided, without alteration."[5]

In England, Priestley continued to look into the strange properties of mercury calx and applied the same goodness test to the air at about the same time that Lavoisier did. But instead of stopping when the contraction of the test air reached the "goodness" of common air, he added still more nitrous air and saw the volume of the test air shrink further. "Applying this test," wrote Priestley,

> I found, to my great surprize, that a quantity of this air required about five times as much nitrous air to saturate it, as common air requires. . . . A candle burned in this air with an amazing strength of flame; and a bit of red hot wood crackled and burned with a prodigious rapidity, exhibiting

Fig. 4.6. Experiments on respiration. The illustration shows Lavoisier measuring the oxygen consumed and the fixed air (carbon dioxide) given off in human respiration. The man behind the mask is Lavoisier's collaborator Armand Seguin. Madame Lavoisier is seated at the table taking notes. She made this drawing of the experiment, which was performed in 1790 and 1791. *Sources:* Edouard Grimaux, *Lavoisier,* 3d ed. (Paris, 1899). Courtesy of the University of Washington Libraries.

an appearance something like that of iron glowing with a white heat, and throwing out sparks in all directions. But to complete the proof of the superior quality of this air, I introduced a mouse into it; and in a quantity in which, had it been common air, it would have died in about a quarter of an hour, it lived, at two different times, a whole hour, and was taken out quite vigorous.[6]

Priestley, a believer in the phlogiston theory, assumed that this exceptionally good air was common air from which the phlogiston had been removed. Obviously, common atmospheric air contained phlogiston, which poured into it from open fires and from the respiration of animals. (Priestley had already demonstrated that plants revivified air that had lost its "goodness.") He therefore named this new air, quite appropriately, "dephlogisticated air" – air from which phlogiston had been removed. He also had the pleasure of correcting Lavoisier's mistake, adding that, "as a concurrence of unforseen and undesigned circumstances has favoured me in this inquiry, a like happy concurrence may favour Mr. Lavoisier in another; and as, in this case, truth has been the means of leading him into error, error may, in its turn, lead him to truth."[7]

Priestley believed that experimental philosophy progressed by fortunate accident and not by preconceived design. But that was not Lavoisier's way. With this last bit of information Lavoisier could see how the theory of the vaporous state explained combustion. Only a part of atmospheric air, only the "eminently respirable" part, was absorbed by metals on calcination. Atmospheric air had, then, to be a mixture of different chemical substances in the vaporous state. Lavoisier was prepared for this conclusion. In April 1773 he had written: "It is possible, . . . indeed probable, that the air is composed of several vaporous fluids mixed together. For this to happen it is only necessary that there be several fluids on the planet that we inhabit that are volatile enough so that they cannot support the degree of heat of our atmosphere without entering into a state of expansion."[8]

The Rationalization of Chemistry

With the theory of combustion falling into place, Lavoisier was eager to capitalize on his victory by applying his new ideas to the explanation of other chemical phenomena, especially acidity. In a series of papers culminating in his "Considérations générales sur la nature des acides" ["General considerations on the nature of acids"] of November 1778, Lavoisier accumulated evidence to demonstrate that the eminently respirable air that was responsible for combustion was also the source of acidity. The next year he was convinced enough of his theory to propose a new name, "oxygen," meaning "acid former," in place of "eminently respirable air." He admitted that he had not been able to release oxygen from marine acid (hydrochloric acid) but assumed that this recalcitrance on the part of marine acid was not sufficient to invalidate a general rule. Marine acid would also yield to analysis in time. It did, but it yielded to Priestley's analysis, not Lavoisier's, and it turned out to be the exception that disproved the rule. Marine acid contains no oxygen. It is the hydrogen ion, not oxygen, that is the principle of acidity.

Another of Lavoisier's puzzles was solved for him when Henry Cavendish found the product of the combustion of "inflammable air." This air (hydrogen) burned in the atmosphere but seemed to leave no product. This was not a problem for the phlogiston theory, because inflammable air, which was regarded as pure phlogiston, should dissipate completely into the air. According to Lavoisier's oxygen theory, however, it should combine with oxygen to form a product, and no product could be found. In 1781 Priestley used an electric spark to explode inflammable air with common air in a closed container. He noticed that the inside of the glass container became dewy after the explosion. His friend John Warltire repeated the experiments in copper vessels and reported smoke as well as dew.

These experiments had been undertaken to measure the weight of heat, rather than the product of burning inflammable air, and therefore Priestley and Warltire mentioned the "smoke" and "dew" only incidentally and emphasized an apparent loss of weight in the experiment. Cavendish heard of these results from Priestley and repeated the experiment in 1781 with a much more elaborate apparatus that allowed him to collect a sizable amount of "dew." It had "no taste nor smell, . . . left no sensible sediment when evaporated to dryness; neither did it yield any pungent smell during the evaporation; in short, it seemed pure water."[9] Cavendish experimented only for his own illumination and thought little about illuminating others. Therefore he was always slow to publish and did not read his memoir on this subject to the Royal Society until January 15, 1784. (The memoir contained much more than his analysis of the "dew.")

In June 1783, Cavendish's assistant Charles Blagden (1748–1820) was in Paris and described Cavendish's experiments to Lavoisier. For Lavoisier this was heaven-sent news. It meant that his theory not only explained the combustion of inflammable air but also indicated that water, which had always been considered an element, was in fact a compound of inflammable air and oxygen. Lavoisier hastily confirmed Cavendish's experiment and reported his results immediately to the Paris Academy of Science. It does not speak well for Lavoisier that he barely mentioned Cavendish in the full memoir that he read to the academy in November 1783. Lavoisier may have felt that his own additional experiments analyzing water by the rusting of iron and his theoretical explanation of the combustion of inflammable air overshadowed Cavendish's experiments, but his slighting of Cavendish was certainly less than generous.

The analysis of water puts the capstone on the oxygen theory. Lavoisier had already shown air to be a mixture; now he had shown water to be a compound. Two of the ancient elements had succumbed to his analysis in ten years. In the language of the new chemistry, inflammable air became "hydrogen," meaning "water former." The new names consolidated the new theory.

The rest of Lavoisier's Chemical Revolution had a more polemical character. In the 1780s several prominent French chemists publicly accepted Lavoisier's theory, and in Edinburgh Joseph Black also began teaching the new chemistry. In 1783 Lavoisier carefully analyzed the phlogiston theory in his "Reflexions sur le phlogistique" ["Reflections on phlogiston"]. He was able to show that the use of the phlogiston theory to explain experiments resulted in many contradictions and inconsistencies. Phlogiston had meant different things to different chemists and had changed its meaning from circumstance to circumstance. Sometimes it was a substance, sometimes a principle; sometimes it had weight, sometimes not; sometimes it was heat, sometimes fire, sometimes the principle of fire. With a logician's attention to the demands of consistency, Lavoisier showed that the phlogiston theory could not meet the requirements of a quantitative chemistry.

In its place Lavoisier, with the assistance of Guyton de Morveau, Antoine-François de Fourcroy (1755–1809), and Claude-Louis Berthollet (1748–1822), began to work out a new language for chemistry. Many chemical names had come from alchemy and were intentionally obscure (vitriol of Venus, green lion, stellate regulus of Mercury); others were named after their discoverer (Glauber's salt, Kunckel's phosphorus, fuming liquor of Libavius); others were named after their place of origin (Epsom salts); and still others after their physical characteristics (butter of antimony, liver of sulfur). The aim of the new chemical nomenclature was to make the names of compounds describe their composition.

The reform of language was a major theme of the Enlightenment. Language was not only a set of symbols but also a mode of reasoning. If the symbols of language were clear and precise and if the grammar were logically sound, then speaking correctly would be the same thing as reasoning correctly. Abbé Condillac claimed that algebra was the best language because it had the best symbols. There was no ambiguity in their meaning, and the grammar of this "language" was such that conclusions followed absolutely rigorously from premises. Lavoisier used this argument to introduce his *Méthode de nomenclature chimique* [*Method of chemical nomenclature*] (1787). He asserted that

> A well-composed language, adapted to the natural and successive order of ideas will bring in its train a necessary and immediate revolution in the method of teaching and will not allow teachers of chemistry to deviate from the course of Nature; either they must reject the nomenclature or they must irresistibly follow the course marked out by it. The logic of the sciences is thus essentially dependent on their language.[10]

It is hardly necessary to add that the new language was to be the language of Lavoisier's chemistry.

As Black noted, acceptance of Lavoisier's nomenclature meant acceptance of the oxygen theory. Acids were named after the amount of oxygen they contained. Thus *sulfuric* acid contained more oxygen than *sulfurous* acid. Its salts were *sulfates* or *sulfites,* depending on the amount of oxygen they contained. The calxes were "oxides." "Mercuric oxide" could mean only the metal mercury combined with oxygen. Once the principles of the nomenclature were understood, there was very little room for confusion. Lavoisier's nomenclature seems obvious to us because it is familiar. It is the system that chemists still use.

Finally, in 1789, Lavoisier wrote a textbook, *Traité élémentaire de chimie* [*Elementary treatise of chemistry*], that placed the oxygen theory and the new nomenclature at the center of a complete revolution in chemistry. Future chemists seeking to learn about his new discoveries were confronted not with random experiments but with a compelling argument leading to a complete system of chemistry.

The organization and rationalization of chemistry was Lavoisier's greatest achievement. The new nomenclature and the new *Traité* purged chemistry of its last alchemical associations. The essences and influences were largely gone, to be replaced by a strict quantitative accounting of chemical substances and their combinations. The discovery of oxygen, however, which is often credited to Lavoisier, was not all his work. In fact it is almost impossible to know exactly what one means by "discovery" in this case. Priestley was the first to recognize the chemical properties of the new "air," but he understood it only in terms of the phlogiston theory. In Sweden, Carl Wilhelm Scheele (1742–86) produced oxygen completely independently from Priestley and Lavoisier but did not publish any of his experiments until his *Chemische Abhandlung von der Luft und dem Feuer* [*Chemical treatise on air and fire*] of 1777. He heated sulfuric acid with "manganese" (manganese dioxide) and collected a gas in which a candle burned brightly. He called it "fire air" and recognized that it was part of the atmosphere, the other part being "vitiated air." Scheele also adhered to a phlogiston theory of combustion, which meant that he did not understand combustion with the clarity of Lavoisier, but he was very close to the idea that "fire air" was a separate chemical entity and a component of the atmosphere.

The fact that oxygen was being identified independently by researchers all over Europe in the 1770s means that it would certainly have been "discovered" without Lavoisier's assistance and that it would also have been recognized as a component of the atmosphere, whatever theoretical interpretation might have been given to its role in combustion. The conceptual foundation of the Chemical Revolution was not just the theory of combustion but the whole theory of the gaseous or vaporous state. Lavoisier began his *Traité* with a long discussion of the term *caloric,* his name for the substance or cause of heat. It was this "caloric" that combined with chemical substances to change them to a vaporous or aeriform state. Lavoisier chose the word *gas,* a term coined by Johannes Baptista van Helmont (1579–1644) in the seventeenth century and reintroduced by Macquer to describe any chemical in the vaporous state. Oxygen, hydrogen, and nitrogen were all different "gases." Lavoisier redefined the word *air* to mean only the atmosphere, which he understood to be a mixture of "gases."

Chemical Atomism

Lavoisier was not the only one to reap chemical rewards from the physical theory of the gaseous state. The reintroduction of atomism into chemistry was accomplished by a meterologist, John Dalton (1766–1844), who became a chemist only when he saw the implications for chemistry of his ideas about the atmosphere. If the atmosphere is a mixture of gases as the new chemistry had demonstrated, why, Dalton wondered, don't the different gases separate into layers under the force of gravity, with the densest gas at the bottom of the atmosphere and the least dense at the top? He also wondered how one could explain the direct proportion between the pressure of a gas and its solubility in water, or the fact that the pressures and solubilities of gases are not changed when other gases are mixed in with them. These were all appropriate questions for a meterologist to ask; they all led Dalton to the conclusion that the atoms of gases in the atmosphere repel other atoms of their own kind but completely ignore atoms of another kind. The atoms of different gases must be physically different, probably in weight and size, in order to act only on atoms of the same kind. Moreover, why are dense gases more soluble than light ones? Something about the weight of the atom must determine the case with which it is held in solution. Atoms of different gases were obviously different, and if one would tell something about their relative weights, one might be able to understand how they interacted in a mixture of gases in contact with a solvent such as water.

The place to look for relative weights was in chemistry. In France, Joseph-Louis Proust (1754–1826) and Claude Berthollet were debating whether chemical substances combined in a definite proportion by weight or in a range of proportions, depending on physical conditions. Proust said the proportion of the reactants was always the same; Berthollet said that it depended on the temperature, pressure, and concentration of the reactants. Proust was right, but not obviously so, because many substances that we now know to be mixtures, such as glasses and amalgams, appeared then to be compounds. Also, many substances combine in several different proportions (the oxides of copper and carbon, for example), which gives the impression that their components combine in a range of ratios by weight. In spite of these doubts, the "law of definite proportions" gained favor with chemists, and for Dalton it was the obvious way to judge the relative weights of atoms of different gases. If the atoms of two substances combined in a definite pattern to make molecules of the compound, and if all molecules of the compound were the same, then that pattern gave the ratio of the numbers of atoms of the two substances entering the molecule. The same ratio would hold for any number of molecules. Chemists could easily weigh the reagents to obtain the ratios by weight of the substances entering chemical combination. Then, if they were able to discover the pattern of the molecule, they could determine the relative weights of the atoms. Of course Dalton did not know the patterns, but in his *New system of chemical philosophy* (1808) he assumed them to be simple and therefore guessed correctly much of the time. At a time when ratio and proportion were much more a part of elementary mathematics than they are now, this all must have been more obvious. But it had not been obvious to the chemists, who had been frustrated throughout the eighteenth century in their efforts to explain chemistry in terms of attractions between the atoms.

The history of the theory of valency and the invention of the periodic table is beyond the scope of this volume, but it is worth noting that chemical atomism, the idea of which could have come logically from the debates of chemists over the law of definite proportions, did not come from them directly but required further inspiration from the theory of heat and of the gaseous state. When Lavoisier sent his *Traité élémentaire de chimie* to Benjamin Franklin in 1790, he wrote that he had avoided theory, "in order to follow as much as possible the torch of observation and experiment. This course, which had not yet been followed in chemistry, led me to plan my book according to an absolutely new scheme, and chemistry has been brought much closer than heretofore to experimental physics."[11] By bringing the physical theory of the gaseous state into chemistry, Lavoisier and his associates believed that they were bringing chemistry and physics together, but from our perspective we would have to say that they made chemistry more *chemical.* The three Aristotelian elements that represented the three physical states of matter – earth, water, and air – had been demolished by the Chemical Revolution. Water and air were found to be a compound and a mixture, respectively, of substances known by their chemical properties. Earth had been resolved into a variety of "earths," all of which would become liquid or aeriform if brought to a sufficiently high temperature. Lavoisier's elements were those simple substances that could be analyzed no further. Unlike Aristotle's elements, they were known by chemical, not physical, characteristics. Of the four Aristotelian elements, only fire remained a puzzle. Lavoisier felt more comfortable with it when he had given it the new name *caloric,* but it was still an anomaly in his list of simple substances. It still differed from the chemical elements in its weightlessness and in its ability to pass through the walls of containers. The final elimination of fire as a substance had to wait for thermodynamics and the kinetic theory of the nineteenth century.

CHAPTER V

Natural History and Physiology

This chapter is about the world of living things. It could be called a chapter on biology, except for the fact that *biology,* as a word and as a discipline, did not appear until the very end of the eighteenth century. To see the world the way the men and women of the Enlightment saw it, we have to see it through the eyes of natural history. "Natural history" means an inquiry or investigation into nature; and "nature," in the Aristotelian sense, means that part of the physical world that is formed and that functions without the artifice of man. A growing tree and a falling rock are both part of "nature" because they move and grow without human direction. Natural history, then, covers the entire range of observable forms from minerals to man, excluding only those objects crafted by human hands and intelligence. Its method is descriptive, and its scope is encyclopedic. Francis Bacon called it the "great root and mother" of all the sciences and made it the indispensable prelude to his experimental philosophy.

In spite of its enormous scope, natural history did not treat all questions about living things. The purpose of natural history was to describe and classify the forms of nature; it did not include a search for causes. Both plant and animal physiology – that is, the investigation of plant and animal functions as opposed to their forms – were still part of physics. When the Paris Academy of Sciences was reorganized in 1699, the sections that dealt specifically with living beings were the descriptive sciences of botany and anatomy. Any experimental physiology that took place was done in the physics section. In the *Encyclopédie,* all history, including natural history, was classified under the faculty of memory; physics, which included zoology, botany, and medicine, was classified under the faculty of reason. Natural history and physiology were separated by the different methods that they followed and by the different goals that they pursued.

Medical doctors dominated the study of living things because they were the only ones to receive formal instruction on such matters. All members of the botany and anatomy sections of the French Academy were doctors. The naturalists Joseph Pitton de Tournefort (1656–1708), Antoine de Jussieu (1686–1758), and Carl Linnaeus (1707–78) were also medical men. So were Bernard de Jussieu, Hermann Boerhaave, and Georg Stahl, whose names we have already encountered in the chapters on experimental physics and chemistry. In the eighteenth century, however, medical training did not lead inevitably to medical practice. Botany, in particular, was a subject that was beginning to be pursued for its own sake, independent of the needs of pharmacy. Just as medicine was losing its dominance over chemistry in the eighteenth century, so did it lose its dominance over physiology and natural history.

As the century progressed, more of the major contributors to natural history and physiology were persons without medical training. The most striking example is in chemistry. Stephen Hales, Joseph Priestley, and Antoine Lavoisier, while they were bringing about a revolution in chemistry, also clarified the functions of respiration, animal heat, and the relationship between plants and the atmosphere. None of these men were doctors, nor were Charles Bonnet, Abraham Trembley, René de Réaumur, Comte de Buffon, Lazzaro Spallanzani, and Jean Lamarck, whose names will figure prominently in this chapter. (The situation in Scotland was different: William Cullen and Joseph Black were professors of both medicine and chemistry.) Through the academies of science it was possible to obtain positions of status without medical training, and as a result natural history and physiology were no longer the exclusive province of doctors or limited to the subject matter of medicine.

The Mechanical Philosophy and the Study of Life

Descartes had concluded in 1638 that with the exception of the human rational soul all natural objects were caused by inert particles of matter in motion. There was, for him, no basic difference between one's watch and one's pet dog. Of course not all mechanical philosophers were so extreme in their views; most were unwilling to carry their principles this far. Robert Boyle, for instance, distinguished between the watch as a work of man and the dog as a work of God, but nevertheless he retained the same mechanical view of nature as Decartes. From this point of view there was no essential difference between living and nonliving objects. Animals were automata; some philosophers even attempted to build robots that would simulate vital functions. The mechanical philosophy swept physiology in the wake of Descartes, so that by 1670 all major physiologists were mechanists. The *De motu animalium* [*On the motion of animals*] (1676), by Giovanni Alfonso Borelli (1608–78), is the most famous example. Borelli analyzed the mechanics of the muscles and skeleton of the human body and tried to explain muscular contraction as a hydraulic or mechanical inflation of the tissue (see Figure 5.1). He also measured the force of muscle, with special attention to digestion in the stomach, which he believed to be primarily a crushing and grinding process.

The mechanical philosophy was seductive, but it could explain vital phenomena such as growth, nutrition, and reproduction only by resorting to the most outlandish hypotheses, none of which was confirmable by experiment. During the first half of the eighteenth century this easy optimism waned, and physiologists realized that a mechanical analysis of living things might be impossible. In 1733 Bernard Fontenelle stated that mathematics certainly did apply to living things but had been unsuccessful in explaining how they functioned because of their great complexity. Life may be merely the result of mechanical organization, but if so it is beyond the reach of investigation. A better alternative would be to study the vital phenomena themselves and attempt to reduce them to rule, without any suppositions about original causes or imagined mechanisms. As a result, experimental physiology in the eighteenth century became phenomenalistic. Experimenters described and linked vital phenomena to the best of their ability without attempting mechanical models.

Natural history experienced a rebirth in the late seventeenth century at the time when the mechanical philosophy was most strongly held. There were several reasons for the new enthusiasm for natural history. One was religious. The mechanical philosophy recognized a creator God but denied him any role in everyday operations of the universe. Therefore God could be known in nature not from any acts but only from the extraordinary complexity and harmony of his creation. Natural history described this complexity in great detail.

From the beginning of the eighteenth century, English natural philosophers published a continuous stream of books designed to reveal the wonders of God's creation through the new sciences. The *Cosmologia sacra* [*Sacred cosmology*] (1701) of the microscopist and plant physiologist Nehemiah Grew (1641–1712) was followed

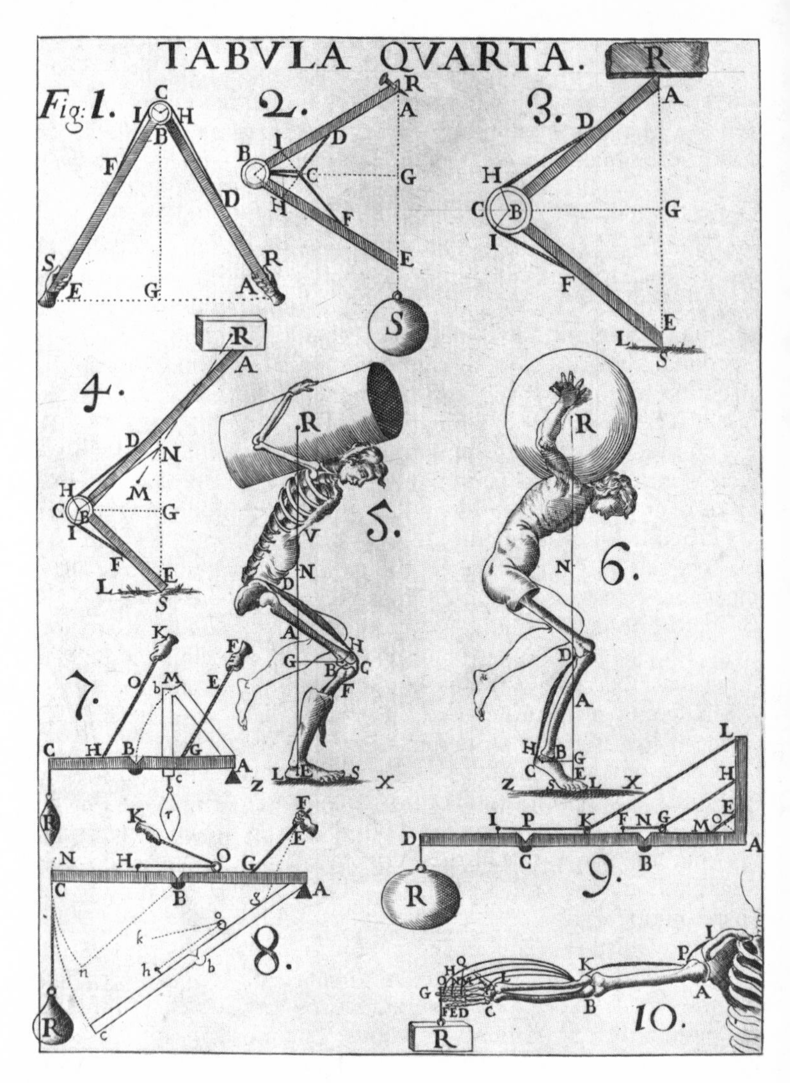

Fig. 5.1. Physiology as mechanical philosophy. G. A. Borelli, in his *De motu animalium* [*On the motion of animals*] (1680) studied the human body as a machine. The mechanical philosophy could explain the action of the bones and muscles in the limbs, and it could explain the action of the heart, but beyond these most obvious mechanical functions it failed to explain the vital functions of the human body. *Sources:* G. A. Borelli, *De motu animalium* (1680). By permission of the Syndics of Cambridge University Library.

in 1704 by John Ray's *The wisdom of God manifested in the works of the creation.* Ray (1627–1705) was the leader of the natural history revival in England and the best naturalist of his age. There was nothing frivolous or shallow about the science that he brought to the revelation of God. The following year, George Cheyne (1671–1743) published his *Philosophical principles of religion natural and revealed,* and in 1713 William Derham (1657–1735) published *Physico-theology, or a demonstration of the being and attributes of God from his works of creation* (see Figure 5.2). *Physico-theology* gave the name to this genre of natural history, and Derham soon followed it with an *Astro-theology.* Ray's *Wisdom of God* went through six editions in ten years, and Derham's *Physico-theology* had three London editions in one year and five French editions. The enthusiasm for physico-theology reached the Continent in the 1720s, where the most prominent contribution was the enormously popular *Le spectacle de la nature* [*Spectacle of nature*] of Noël-Antoine Pluche (1688–1761), which began to be published in 1732 and reached its eighth volume in 1750. In France, natural theology declined after 1750 as a result of the antireligious sentiment of the Enlightenment, but in England it continued well into the nineteenth century, where it finally encountered its nemesis in Charles Darwin (1809–82).

A second reason for the success of natural history was a desire to get rid of the animistic "principles" and "souls" that had characterized Renaissance science. Natural history described and classified all three kingdoms of nature – animal, vegetable, and mineral. As a science of forms and categories, it did not have to concern itself with the causes of life and therefore could easily include living and nonliving things within the same schema. Natural history was a complement to the mechanical philosophy, because both approaches to the natural world merged the living and the nonliving together. There was no room in either approach for spirits other than the human rational soul. This characteristic of natural history in the early Enlightenment has led several modern philosophers and historians, especially Michel Foucault, to state that there could be no science of biology before 1750 because there was no understanding of life separate from the nonliving world. Because modern biology attempts to explain life in physical-chemical terms, we may be tempted to think of the mechanists of the seventeenth and eighteenth centuries as precursors of the modern view. In fact the creation of biology as a separate discipline came only after a strong reaction against the mechanical philosophy had separated the study of living things from inanimate nature and had explained "life" by principles that did not apply to the inanimate world.

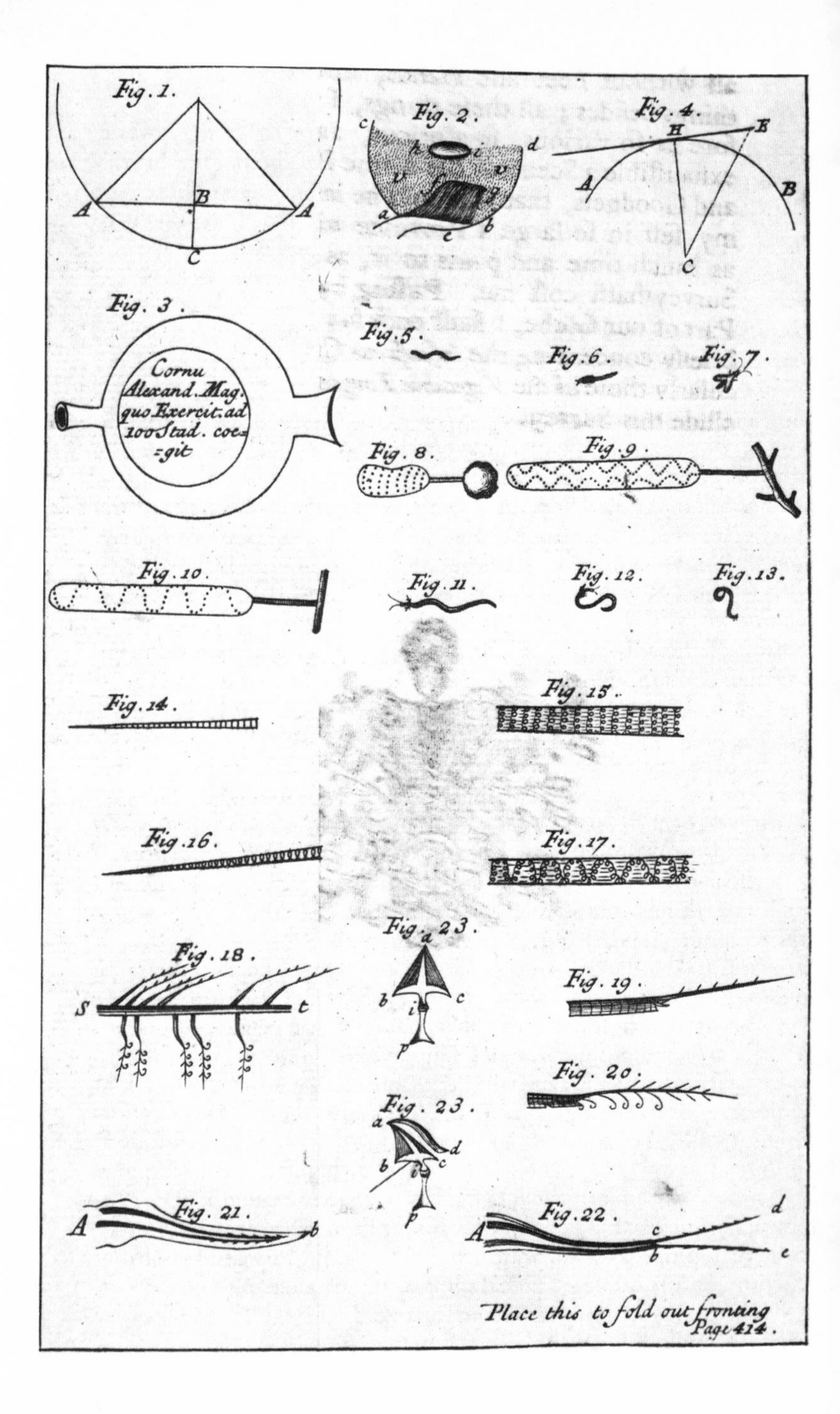

Place this to fold out fronting Page 414.

A third reason for the rise of natural history in the 1670s was the increased emphasis, especially in England, on the empirical sciences. Without repudiating the mechanical philosophy, British philosophers and scientists did repudiate Descartes's rationalist, a priori approach to the study of nature. The world is known from careful observation and study of natural phenomena, not from deductive reasoning on abstract principles. It is no coincidence that natural history revived first in England, where the experimental tradition was the strongest, and only then passed on to the Continent.

Experimental Physiology

The rise of experimental physiology in the 1740s coincided with the appearance of the theory of subtle fluids in experimental physics and chemistry. Just as chemistry turned from theories based on attractions and repulsions between the atoms to a study of the chemical properties of acidity, alkalinity, and metallicity, so physiology turned from the description of the body's organs as levers, pulleys, pumps, and sieves to an investigation of those characteristics such as growth, nutrition, and regeneration that make living

---

Fig. 5.2. Examples of God's design. William Derham, in his *Physico-theology, or a demonstration of the being and attributes of God from his works of creation* (1713) selected these "contrivances" to illustrate God's wisdom. Derham was especially impressed by the clothing of animals. God created man naked because man was endowed with the faculty of reason, which meant that he would be able to help himself, "but for the poor shiftless Irrationals, it is a prodigious Act of the Great Creator's Indulgence, that they are already furnished with such clothing as is proper to their Place and Business." Figures 14 through 17 depict the appearance of mouse hair, as seen under a microscope. Derham thought that the spiral lines on the hairs might aid the "insensible perspiration" of the mouse. Figures 18 through 20 are microscopic views of bird feathers. God has designed them so that they will interlock and grasp each other, thereby giving the birds "an easy Passage through the Air" and assist "in wafting their Body through that thin Medium." Figures 21 and 22 show the stinger of a wasp, "so pretty a piece of Work, that it is worth taking Notice of," and Figure 23 shows the inner ear of a bird. All of these contrivances reveal God's concern for the welfare of his creatures and prove his wisdom and beneficence. *Sources:* William Derham, *Physico-theology: or, A demonstration of the being and attributes of God, from his works of creation,* 2d ed. (London, 1714), p. 414. Courtesy of the Rare Book Collection, Special Collections Division, University of Washington Libraries.

things different from machines. Before 1740, the standard authorities in physiology were Giovanni Borelli, Lorenzo Bellini (1643–1704), Archibald Pitcairne (1652–1713), and James Keill (1673–1719) – all mechanists – and the mechanist Boerhaave was the major authority in chemistry. After the middle of the century, the Germans Stahl, Friedrich Hoffmann (1660–1742), and Albrecht von Haller became the major authorities in physiology, and Stahl, Etienne Geoffroy (1672–1731), and Joseph Macquer were the major authorities in chemistry. It is tempting to describe this shift as a change from mechanism to vitalism, but creating such absolute dichotomies in the history of science always gets us into trouble, because it ignores the middle ground and the fact that most scientists are more interested in their experimental results than in global theories such as mechanism and vitalism. The comparison to chemistry is again helpful. The new principles in chemistry and physiology were meant to stand for observable qualities, not for the old imagined "souls" or "influences" of Renaissance animism. When vitalism revived around 1760, it was in a strictly experimental context. The failure of mechanical theories made physiology more phenomenalistic.

James Parsons (1705–70) noted in 1752 that mechanical philosophers had sought in vain for "Particles and Pores of different configurations, in vain had Recourse to the *Momentum* of the Blood and in vain endeavoured to reconcile the Doctrine of Secretions to Mathematical Calculation,"[1] and at the same time Diderot argued that the taste of the times had turned to chemistry and physiology because those sciences dealt with nature as it existed, rather than with nature as a mechanical and mathematical abstraction.

Of course the new science of experimental physics and chemistry had a direct influence on physiology and medicine. Electricity promised to hold answers for physiology. The electric eel and the sensitive plant were candidates for study because they both appeared to protect themselves electrically. Electricians in England, France, and Germany concluded from their experiments that electrified seeds germinated faster, that electrified plants sent out shoots earlier, and that electrified animals were slightly lighter than nonelectrified ones. The electric torpedo fish, the electric eel of the Guianas, and the electric catfish of Africa were all studied to discover the source of their electricity. It was difficult to explain how these animals could produce shocks while immersed in a conducting medium, but Henry Cavendish showed that given large enough capacitance a shock could be delivered under water. He even built a model torpedo fish out of leather attached to a large Leyden jar to prove his point.

Electricity produced muscle contractions, which indicated that electricity in the body probably took the form of a fluid that moved through the nerves carrying sense stimuli and motor commands. But the failure of the earlier mechanical theories urged caution, and Haller, the leading physiologist at mid-century, argued that it would be premature to identify the electrical material with animal spirits. Haller's caution was wise, because experimental technique in the eighteenth century was totally incapable of revealing the electrochemical nature of nerve impulses. Medical doctors, however, were soon using electrical therapy with apparent success. It is not surprising, then, that Luigi Galvani (as we saw in Chapter IV) believed that his frogs' legs contained within them organic Leyden jars that caused the legs to kick when they discharged. The complexity of electrical phenomena in physiology put them beyond the reach of eighteenth-century experimenters, but it is significant that they attempted to apply the results of physical experiments in the world of living things.

The chemists had greater success than the electricians in bringing their skills to the aid of physiology. They could not answer the most important chemical questions such as those concerning the nature of digestion, but their achievements in pneumatic chemistry clarified the relationship between plants and the atmosphere and the production of animal heat. Nehemiah Grew and Marcello Malpighi (1628–94) had both observed pores (stomata) in the underside of leaves and had concluded that leaves used them either to take in air or to exude sap. Earlier in the century Johannes van Helmont had performed a famous experiment in which he grew a willow tree in a carefully weighed amount of soil. Since very little if any soil was consumed, he concluded that the increased matter of the tree came from the water that he had regularly added to the soil. Stephen Hales, that intrepid searcher for air, placed a peppermint plant over water under a glass cylinder and discovered that some air seemed to be consumed by the plant, while Priestley, who knew that air came in different varieties, found in 1772 that a mint plant would revivify air in which it was grown. The experiments were difficult to duplicate because Preistley did not fully recognize the importance of light for the action of the leaves and because he did not believe that the scummy "green matter" (algae) that covered the inside of his glass vessels was also a plant. He did, however, collect some bubbles from the leaves in 1778 and found the air to be "dephlogisticated air" (oxygen). Most important, from his point of view, was the discovery that plants are responsible for revivifying the air that combustion and the respiration of animals are constantly polluting with phlogiston. This aerial balance between plants and animals was to him another example of the harmony of God's creation.

Priestley's successes inspired the Dutch physiologist Jan Ingen-Housz (1730–99) to take up the problem. Ingen-Housz was able to show in his *Experiments on vegetables* (1779) that it was sunlight, not heat, that was essential for the production of oxygen by the leaves. He found that in the dark the leaves reversed this process and emitted small quantities of "fixed air" (carbon dioxide), whereas in sunlight they produced large quantities of oxygen. He observed that only the green parts of plants produced oxygen and that it was emitted from the underside of the leaves. These experiments were done by placing the plant entirely under water but still in sunlight and observing the bubbles of oxygen appearing on the underside of the leaves. Ingen-Housz did find, however, that the leaves had to be placed in fresh pump water, not boiled water, in order to have any oxygen released. He interpreted this experiment to mean that the boiled water absorbed the oxygen from the leaves whereas the pump water, which was already saturated with oxygen, allowed it to escape to the surface.

Jean Senébier (1742–1809) found that Ingen-Housz's explanation could not be correct because atmospheric air was not readily soluble in water. "Fixed air," however was highly soluble in water. Even though fixed air made up only a small fraction of the atmosphere, there was enough of it dissolved in pump water to supply the leaves, whereas boiled water contained little fixed air. Senébier showed that in sunlight the leaves absorbed fixed air and emitted oxygen.

Senébier also showed that it was not necessary to have the entire plant in order to produce oxygen. Just the green leaves, or even chopped leaves, would convert fixed air into oxygen. He even dissected the leaf further and found that it was the green interior of the leaf, the parenchyma, that was responsible for the production of oxygen. This description of Senébier's efforts is anachronistic, in that Senébier interpreted the conversion in terms of phlogiston. But Claude Berthollet soon explained Senébier's results according to the oxygen theory (1788).

Nicolas Théodore de Saussure (1767–1845) amplified the theory in much greater detail. He showed that water was a nutrient of plants and not just a carrier of other nutrients. He also showed that plants can survive in a vacuum and in an atmosphere of nitrogen by secreting small amounts of fixed air and oxygen. If these essential gases are removed, however, the plants die. He found that plants grow better in an atmosphere rich in fixed air, up to a concentration of approximately 8 percent, but that in higher concentrations the plant cannot function. And, most surprising of all, he found that even though plants can grow in an atmosphere that is mostly nitrogen, the nitrogen that they absorb comes from the soil. The realization that "airs" are chemical substances, that they can become "fixed" in plants, and that the atmosphere is a mixture of such "airs" were all essential for an understanding of plant nutrition.

Digestion was another physiological process that appeared to be chemical. Borelli and the iatromechanists (those who tried to explain living things mechanically) postulated that digestion was a grinding and crushing process and found some confirmation from crushed objects found in the gizzards of fowl, but most physiologists believed that digestion was chemical. Van Helmont had proposed an *archaeus* in the stomach that he believed to be the innermost essence of life and that acted by fermentation. In fact he hypothesized six digestions or "concoctions," all stages in the conversion of food into living flesh.

Eighteenth-century physiologists put aside the *archei* and attempted to do experiments directly on the digestive fluids. Regnier de Graaf (1641–73) had investigated pancreatic juices with limited success. Réaumur had persuaded a chicken to swallow a sponge on the end of a string, which he could then retrieve to obtain a sample of the gastric juices. He also had a pet kite (a kind of hawk) that would swallow and later regurgitate perforated spheres in which he placed a variety of substances to analyze the gastric juices. Lazzaro Spallanzani (1729–99) confirmed Réaumur's experiments and did others on the digestive action of saliva. In order to discover whether the juices in the human stomach were like those in animals, he performed the experiments on himself, swallowing various tubes and bags of samples, in spite of the danger to his alimentary canal. In Edinburgh, Edward Stevens (ca. 1755–1834) did much the same experiments with the help of a human carnival performer who swallowed stones for a living. Stevens placed his specimens in hollow perforated silver spheres that the volunteer would swallow and later regurgitate. The state of chemistry in the eighteenth century was not advanced enough to allow a very thorough analysis of digestion, but it is significant that these scientists put aside the vital principles and *archei* and sought a direct experimental analysis of the process of digestion.

Although the attempts to create a chemical physiology of animals were not very successful, there was a noticeable change in the way in which chemistry and physiology were studied after 1740. Instead of trying to arrive at the structure of living things, physiologists placed greater emphasis on vital function. As we have seen, the same thing was true in chemistry. Attempts to discover the structure of matter gave way to a desire to rationalize chemical processes.

Undoubtedly the most important figure in this transition was Georg Stahl, who was both a chemist and a physician. Not only was he the first to criticize mechanistic explanations, but his books on chemistry and physiology had great influence throughout Europe. In spite of his vast knowledge of chemistry, he denied that it had any connection to medicine. Living matter was entirely separate from nonliving matter because living matter contained an *anima sensitiva* that kept it from corruption. Blood, for instance, immediately putrefies when it is lacking this principle of life. No purely chemical analysis can detect the anima, but it is apparent in all living things. Certain parts of the body such as the heart and the limbs undoubtedly serve a mechanical function, but their mechanical purpose is superficial, and a more penetrating investigation shows that they are quite unlike inanimate matter. All organic forms work towards a final goal, but brute machines work blindly, responding only to the motions communicated between their parts. In his *Theoria medica vera* [*True theory of medicine*] (1708) Stahl argued his medical theories aggressively and soon found himself in conflict with his colleague Friedrich Hoffmann, whose theory of medicine was more mechanical (although Hoffmann believed that animals contained an organizing force acting through the ether that was not present in nonliving objects). Stahl's equally famous *Fundamenta chymiae dogmaticae et experimentalis* [*Foundations of dogmatic and experimental chemistry*] (1723) was translated into English by Peter Shaw (1694–1764) in 1730 and was greatly admired by William Cullen for its thoroughness, although Cullen opposed Stahl's medical theories. Robert Whytt, Cullen's colleague at Edinburgh, was the first in Britain to pick up Stahl's physiology in his *Essay on the vital and other involuntary motions of animals* (1751). Whytt argued that the irritability of living tissue came from a living principle contained in it – not necessarily the anima described by Stahl, which was an extension of the soul, but still a quality or characteristic that was unique to living things.

In France, Stahl's medical theories were popular at the medical school at Montpellier. The appearance of truly vitalistic theories around 1760 was the work of graduates from this school, most notably Henri Fouquet (1727–1806) and Gabriel-François Venel (1723–75) (both of whom wrote articles on physiology and chemistry for the *Encyclopédie*) and Théophile de Bordeu (1722–76), who was chosen by Diderot to be the expert interlocutor in his dialogue entitled *Rêve de d'Alembert* [*D'Alembert's dream*] (1769).

At the center of these new debates in physiology was Albrecht von Haller. Haller was born at Berne and studied at Tübingen and Leiden under Boerhaave. In 1736 he became professor of medicine at the University of Göttingen, and it was there that he carried out his famous investigations into the sensibility and irritability of animal tissue. The property of "irritability" had first been recognized by Francis Glisson (1597–1677), who used it to explain why the gall bladder does not discharge bile into the intestines constantly but only when bile is needed. Haller did experiments to show that the gall bladder discharged more bile when irritated and therefore that irritability performed a controlling function in the body.

Haller generalized this concept of irritability and distinguished it from sensibility, which he believed to be an entirely different property. Irritable tissue contracts when it is touched. Sensible tissue sends a message to the brain. Thus nerve tissue is eminently sensible but not irritable, because it does not contract upon touch. Tendons, bones, the cerebral membrane, liver, spleen, and kidney all lack sensibility. Muscle tissue is sensible, but it is also highly irritable. Although the nerves themselves did not appear to be irritable, Haller showed that the diaphragm could be made to contract by irritating severed nerves, which indicated that nerves had some connection with irritability. But this irritability seemed to be a property of the material of the muscle itself and did not depend on the action of the soul. Stretched muscle fibers contract spontaneously to their former length. Irritability could not be a vital force because it continued for some time after death. Haller's careful experimental technique was matched by the cautiousness of his theorizing. He refused to explain irritability by any abstract and unspecified vital force, nor would he accept a completely mechanical model. He saw his physiology as an *animata anatome,* an experimental science that investigated and explained the special properties and functions of living matter without going beyond the information obtained from the senses.

Bordeu and the other doctors from the medical school at Montpellier criticized the distinction that Haller made between irritability and sensibility. Bordeu claimed that all living matter was sensible and that irritability was only a special case of sensibility. It was this belief in a universal property of sensibility, not any religious commitment to belief in an immortal soul, that made Bordeu a vitalist. Haller was the one whose religious convictions caused him to insist on the unity and spirituality of the soul.

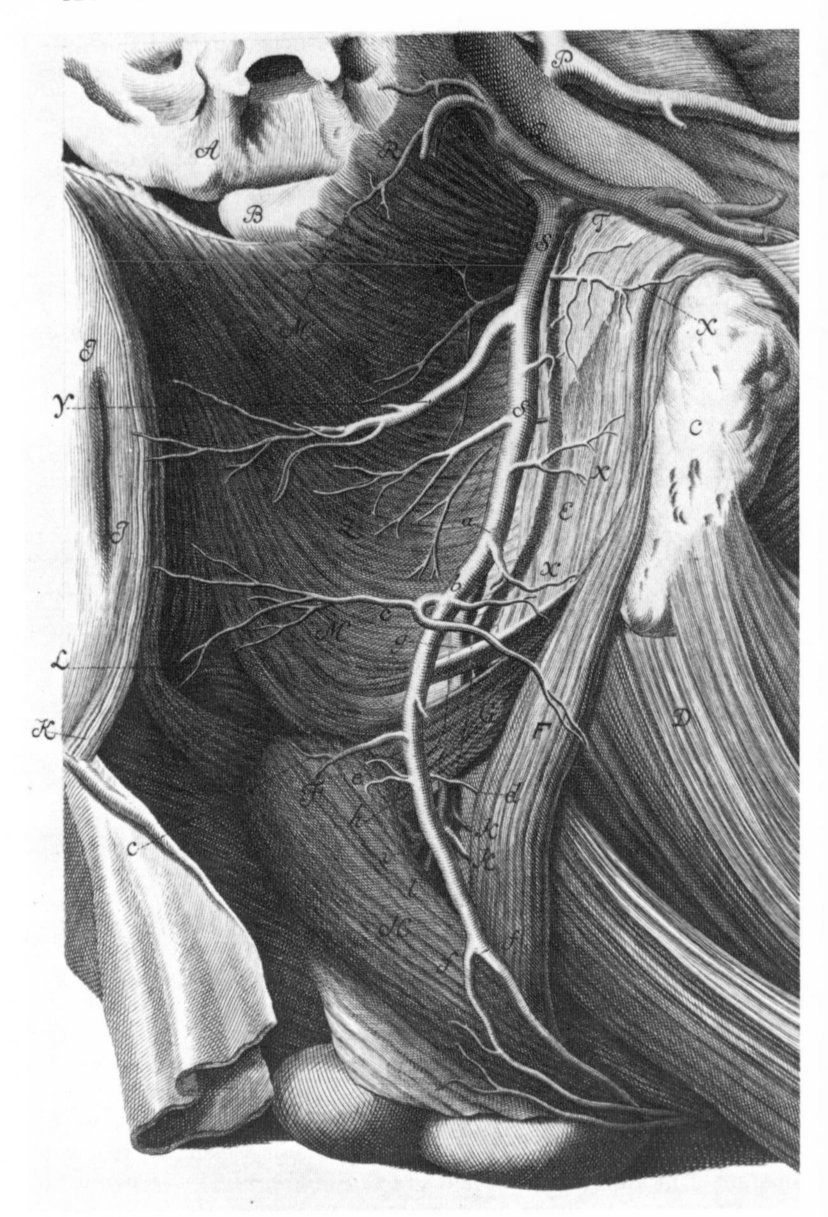
P
A
R
B
S
T
X
M
Y
C
X
E
X
L
D
K
F
d
c
N
f
f

It is here that we confront the central paradox of physiology during the Enlightenment. The antireligious sentiments of the philosophes inclined them to materialism and away from any dependence on the Christian concept of the soul. Mechanism, however, had failed in physiology because it could not account for the properties of life. Therefore animals could not be machines composed of inert particles. The answer of the materialists was to revive the ancient Stoic *pneuma* and endow all of matter, or at least all of organic matter, with life. The Stoics had chosen activity and change, rather than structure and permanence, as the foundation of nature. They explained natural phenomena by forces rather than by the organization of matter. The *pneuma* was the breath of the cosmos, the activating principle responsible for all change and all life. The materialist philosophers of the eighteenth century made matter active by giving it the properties of life. In essence, they distributed the soul throughout matter in order to get rid of it.

Diderot, who carried his materialism as far as the evidence and his common sense would allow, preferred the physiology of Bordeu over that of Haller. In *D'Alembert's dream* the world becomes a living being, infinitely elastic and full of force. Stones become thinking beings, and thinking beings incessantly change their forms. "All beings circulate from one to another; as a result all species . . . are in perpetual flux. . . . All animals are more or less men; all minerals are more or less plants; all plants are more or less animals. Nothing is precise in nature."[2] For Diderot there was no difference between the organic and the inorganic except in the degree of organization. His whole world was dynamic. The universe was a great animal, and it was also one enormous elastic body conserving *vis viva.* There was no real difference in his philosophy between the dynamic and the vital, no difference between physics and physiology.

The philosopher who had argued most compellingly for the existence of force and vitality in matter was Leibniz. But Leibniz would

---

Fig. 5.3 Anatomic illustration in the eighteenth century. Albrecht von Haller was the acknowledged leader in the study of physiology and anatomy during the eighteenth century. This drawing by Joel Paul Kaltenhofer illustrates the major artery in the human pelvis for Haller's anatomic study *Icones anatomicae* (1749). *Sources:* Albrecht von Haller, *Icones anatomicae quibus praecipuae aliquae partes corporis humani delineatae proponunter et arteriarum potissimum historia continetur* . . . (Göttingen, 1749), pt. 4, illustration entitled "Arter Pelvis T.IV." By permission of the Syndics of Cambridge University Library.

have had no sympathy for the imprecise and everchanging world described by Diderot. As a mathematician and rigorous metaphysician, Leibniz believed that the universe in all past, present, and future states followed a "preestablished harmony" laid down by God at the time of creation. This harmony was maintained by the smallest metaphysical units, or "monads," which were endowed with activity, perception, and will. Thus the properties of consciousness existed at the most fundamental level. Mechanical action was merely a phenomenon detected by our senses. It was "real" in the sense that it could be observed, but it was not fundamental. According to Leibniz, there was no way that mechanism alone could create an animal.

Leibniz's influence undoubtedly crept into all vitalistic thought during the eighteenth century, but it appeared most explicitly in the works of his disciple Louis Bourguet (1678–1742). Bourguet's *Lettres philosophiques sur la formation des sels et des crystaux et sur la génération et le mécanisme organique {Philosophical letters on the formation of salts and crystals and on generation and organic mechanism}* (1729) was mechanistic, but it was mechanism with a difference. Bourguet noted that inorganic matter could grow, as in the formation of crystals, but it always grew by accretion of more matter on the outside, repeating the form of the initial crystal. Living things, on the other hand, grew by molecules added throughout their interiors. He called this process *intussusception* and used the term to distinguish organic from inorganic matter.

Bourguet also pointed out another way in which an "organic mechanism" differs from an inorganic or "general mechanism." A crystal is its own mold and merely repeats its form, but in an organic mechanism the molecules are organized according to an internal arrangement that is not a simple mold. The molecules are "accomodated" to the system of the living being and united to its principle monad. Not all molecules are assimilated into an organic mechanism. Only the organic produces the organic, according to Bourguet. Only organic molecules are assimilated by living things, because the distinct qualities of life exist throughout the organism, even at the molecular level. There is a difference between "organic" and "organized." Organization is only an arrangement of molecules. Life cannot be simply a matter of organization, because organization only determines structure. As we saw in the case of Stahl, the characteristics of life depend not on structure but on vital function.

Bourguet's ideas found popular exposition in the writings of Buffon. Buffon wished to avoid the vitalism of Stahl, but he was also aware of the inadequacies of strict mechanism. He retained a belief in atoms but held that living matter was composed of "organic molecules" that the organism took in through nutrition and sifted out from the atoms of inorganic matter. The organic molecules were directed by an "interior mold," and the property of intussusception required a special "penetrating force" that carried the organic molecules to their proper places in the interior mold. Buffon had begun his career as a mathematician and a strong disciple of Newton. (He translated Newton's *Treatise of fluxions* and Hales's *Vegetable staticks* into French.) But he realized, as Leibniz had realized before him, that attraction and repulsion between inert atoms could never make a living animal. Some special force and guiding principle beyond those of mechanics would have to be added.

The most scandalous physiology of all was *L'homme-machine* [*Man the machine*] (1748) of Julien Offroy de La Mettrie. La Mettrie studied with Boerhaave and spent his early years translating the works of his mentor. When he came to create his own theory, it was blatantly materialistic and atheistic. In his human machine there was no essential difference between conscious and unconscious behavior, no freedom of will, no rational soul, and no moral good beyond the perfectibility of the mechanism. La Mettrie drew from Leibniz and cited Haller and the other physiologists of the age, but his polemic sounds much more like the ancient atomists Epicurus and Lucretius than any contemporary author. His radical books were in the vanguard of the antireligious sentiment of the Enlightenment, but he was not in the vanguard of physiology. As speculative philosophy his theory was stimulating and even learned in places, but it lacked the experimental basis that characterized physiology in the 1740s.

Diderot, like Buffon, had studied mathematics and also like Buffon had repudiated it in favor of natural history and chemistry. In fact, Diderot was strongly impressed by Buffon's *Histoire naturelle,* which provided inspiration for his own *De l'interprétation de la nature* [*On the interpretation of nature*] (1753). Diderot's first philosophical work, *Pensées philosophiques* [*Philosophical thoughts*] (1746) had been deistic – that is, it had demonstrated the existence of God through the order of nature and had drawn from the English deists, especially Lord Shaftesbury. But Diderot's *Lettre sur les aveugles* [*Letter on the blind*] (1749) began his gradual conversion to materialism. The *Letter on the blind* discussed the psychology of sensation, and particularly its moral implications. The book was dangerous enough to rouse the authorities, and Diderot was jailed at Vincennes outside of Paris at the crucial time when the *Encyclopédie,* of which he was chief editor, was just getting under way. His *Interpretation of nature* (1753) turned from psychology to scientific method and marked a further step in Diderot's march from a mathematical, rationalist deism to a dynamic, vitalistic materialism.

In his dialogue *D'Alembert's dream* he brought together the ideas of Spinoza, Leibniz, John Toland, Buffon, Maupertuis, Haller, and Bonnet into one wildly speculative account of life. The dialogue begins with a late-evening conversation between Diderot and his friend d'Alembert. Afterward, tired of Diderot's wild speculations about the nature of life, d'Alembert goes off to bed. During the night he begins talking in his sleep about the subjects that he and Diderot had debated earlier. Alarmed by his raving, his mistress, Mademoiselle de Lespinasse, calls Doctor Bordeu to his bedside. Bordeu understands the profound importance of d'Alembert's words and explains them to Mademoiselle de Lespinasse. The dialogue mixes philosophy, science, art, and speculation in a style that is uniquely Diderot's. Diderot claimed that it was at once the craziest and the most profound writing possible. In a later letter of 1765 he described the basic idea of *D'Alembert's dream* and revealed his debt to Bordeu. "Sensibility is a universal property of matter, a property that lies inert in inanimate objects [but one] that becomes active in the same objects by their assimilation into a living animal substance . . . The animal is the laboratory in which sensibility, beginning from its inert state, becomes active."[3] In the works of La Mettrie and Diderot, the scientific debate over vital function moved into the wider domain of Enlightenment philosophy, with all its social, political, moral, and religious implications.

Generation

The change that we have noted in physiological theories around 1740 was caused by the failure of the mechanical philosophy to adequately account for the functioning of living organisms. Nowhere was this more striking than in the problem of "generation," which included both the reproduction of organisms and the regrowth of body parts. One might suppose an equivalence between the structures of living and nonliving things, but there seemed to be no way that mechanism could account for growth and reproduction. The more physiologists learned, the more inadequate mechanical explanations became. Two sensational new discoveries in generation coincided with the change in physiological theories, and as a result the nature of generation became the most exciting problem in the life sciences.

The first was the discovery of parthenogenesis of aphids by Charles Bonnet (1720–93). Antoni van Leeuwenhoek (1632–1723) had observed that the young of the aphid, or plant louse, were present as miniature adults within the parent. This indicated that aphids, unlike most insects, were viviparous and brought forth their young alive rather than from eggs. Still more surprising was the fact that no one could observe any males. Réaumur, who also made a detailed study of insects, suggested that aphids were all female; Leeuwenhoek suggested that they were hermaphrodites carrying the organs of both sexes, but Réaumur denied this, arguing that there was no sign of male organs in any of the individuals he dissected. Other microscopists claim to have observed two distinct sexes and that the female laid eggs like other insects.

In a careful series of experiments beginning in 1740, Bonnet, on Réaumur's suggestion, took up the problem. He raised a newly born female in seclusion and eventually obtained ninety-five young from this single aphid. In another experiment he raised aphids through ten generations with no males present, demonstrating conclusively that aphids reproduced parthenogenetically. Bonnet's results reinforced the ovist view that the embryo of every species was preformed in the mother as a tiny seed and merely grew. In animals that reproduced sexually the role of the male, according to the ovist theory, was merely to initiate growth of the preformed embryo. This theory did not adequately explain the existence of male characteristics in the offspring, but the ovists argued that the semen, by initiating the growth of the preformed embryo and by nourishing it in its early stages, impressed on the embryo the characteristics of the male.

A second startling discovery was made by Abraham Trembley (1710–84). Trembley's subject was the fresh water hydra, or "polyp," as it was called. These small creatures, about a quarter of an inch long, grew on the bottom of lily pads and other aquatic plants. Leeuwenhoek had observed that they reproduced by budding and had assumed them to be plants, but Trembley found on closer observation that they caught food in their tentacles and delivered it to an interior stomach. They also reacted to touch and could move, using a primitive foot (see Figure 5.4). These characteristics made them animals, but of the lowest form, at that point on the scale of living beings where animal forms pass over into plant forms.

In order to discover whether the polyp could regenerate itself, Trembley cut a specimen in two. To his great surprise he saw each piece regenerate a complete polyp. He then cut polyps crosswise, lengthwise, and in different numbers of pieces. Each piece always

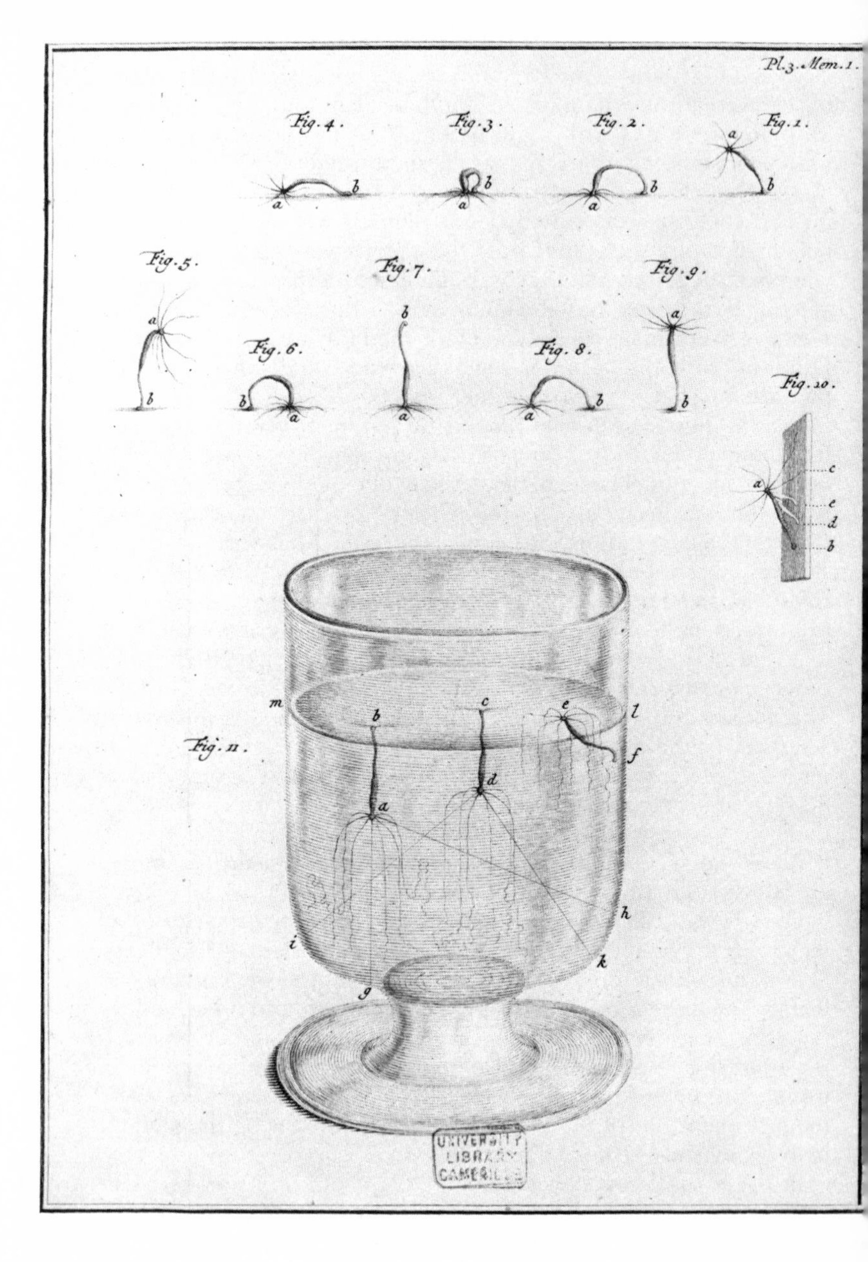
Pl. 3. Mem. 1.
Fig. 4.
Fig. 3.
Fig. 2.
Fig. 1.
Fig. 5.
Fig. 7.
Fig. 9.
Fig. 6.
Fig. 8.
Fig. 10.
Fig. 11.

produced an entire polyp. As an experimental pièce de résistance, he turned a polyp inside out by inserting a bristle into its gut and peeling the body back as one would pull a glove inside out. The polyp accepted its new condition and merely grew an outside on what had formerly been an inside. In 1744 Trembley published a detailed account of his experiments. Réaumur and Bonnet extended them to other animals. They found freshwater worms that regenerated in the same fashion, and since worms were definitely animals, previous doubts about the animality of the polyp lost their force.

These experiments produced a major philosophical dilemma. If each part of an animal could regenerate the entire animal, then where was its "soul," or organizing principle? Naturalists had long known of the ability of crabs and salamanders to regenerate missing parts, but in these cases the severed parts died. It had been assumed that the organizing principle was not in the lost claw or tail but in the animal from which it was taken. In the case of the polyp, however, each piece regenerated and therefore had to contain within itself the power and form needed to reproduce the whole. The modern solution – that animals are composed of tiny cells, each one of which contains within its nucleus sufficient information to create the entire animal – would have appeared ridiculously fanciful at a time when cells were unknown. The microscope had revealed that animals appeared to decrease in size without limit, which suggested that a complex animal might be reduced in scale to a germ or seed, but the seed, in order to become an embryo, would have to be in one part of the animal, not distributed throughout it. To La Mettrie and Diderot, the experiments with the polyp proved that there was no soul and that the properties of life were distributed throughout matter. It was a useful argument for a philosopher advocating materialism and atheism, but it did not help the phys-

---

Fig. 5.4. Trembley's polyp. Is it an animal? And if so, where is its soul? Abraham Trembley found that the polyp could move and feed itself, as shown in this illustration from his *Mémoires pour servir à l'histoire d'un genre de polypes d'eau douce . . .* [*Memoir on the natural history of a species of fresh water polyps*] (1744). This would make it an animal. But it could also regenerate an entirely new polyp from any severed part, which would make it more like a plant. The regeneration of the polyp contradicted the notion that the embryo could be preformed in any part of either parent. *Sources:* Abraham Trembley, *Mémoires pour servir à l'histoire d'un genre de polypes d'eau douce, à bras en forme de cornes* (Leiden; 1744), pl. 3, memoir 1. By permission of the Syndics of Cambridge University Library.

iologist, because it did not explain how this distribution of life took place.

The debate over the polyp complicated a debate that had been going on since the middle of the seventeenth century. William Harvey (1578–1657) had claimed in his *Exercitationes de generatione animalium* [*On the generation of animals*] (1651) that all animals came from an egg. He had experimented with developing chick eggs and with deer from the Royal Park. The "eggs" that he obtained from pregnant deer were actually undeveloped embryos. Harvey did not, however, claim that the embryo was preformed in the egg. Instead he followed the Aristotelian notion that the embryo began as a homogeneous mass and that the organs formed one after another from this homogeneous substance, a process called *epigenesis.*

Epigenesis made sense for an Artistotelian, because according to Aristotle all change takes place by a process in which unformed substance takes on a form that is potentially, but not actually, in it. The mechanical philosophy had repudiated Aristotle, however, along with his concepts of form, potentiality, and final cause. The mechanical philosophy required that the embryo have an immediate mechanical cause. It could not just appear.

In 1688, Jan Swammerdam (1637–80) had shown that the insect larva, pupa, and imago can exist simultaneously, all nested one within the other. He also showed that the legs of a frog were already present under the skin of the tadpole before they began to emerge. Swammerdam concluded from this evidence that there was no epigenesis but that the embryo always existed preformed in the adult. The philosopher Nicolas Malebranche carried preformation to its logical conclusion. Locating the preformed embryo within the parent did not solve the problem of its formation but merely moved it back to the previous generation. Malebranche's solution was to have all generations preformed, one within the next. The seeds of all living things had been formed by God at the Creation and merely unfolded in successive generations. This highly unlikely theory of preexistence solved two difficult problems. It explained the existence of Original Sin, since the entire human race was present in Adam and Eve at the time of the Fall, and it explained where the embryo came from. It also removed the need for the concept of continued spontaneous generation, which was strongly opposed by the mechanical philosophers.

The microscope might have put an end to the theories of preformation and preexistence if it had been powerful enough, but good compound microscopes that made it possible to observe the cell and its structure were not available until the 1830s. Carl Ernst von

Baer (1792–1876) first observed the mammalian egg in 1826, and Wilhelm August Oscar Hertwig (1849–1922) first observed the fecundation of the ovum by sperm in 1875. Embryologists in the eighteenth century could only posit mechanisms of generation to explain what they observed on a grosser level.

The sperm, however, had been observed through the microscope by Leeuwenhoek in 1677. Whereas the ovum had to be imagined, the "animalcules" or "spermatick worms" were plainly visible in the semen. Their mere presence in the semen, however, did not prove that they were the agents of generation. Microscopists regularly found parasites in the blood, intestines, and ovaries of animals, and the spermatozoon could easily be just another parasite that had found its ecological niche in the male testicles. The word *spermatozoon* was coined by von Baer in 1827. The name means "animal in the seed," and it shows that there were able embryologists in the nineteenth century who still believed that the animalcules were only parasites. It was not logical to expect a mammal, for instance, to be fertilized by a worm, especially since the animalcules were never observed in the uterus. It made more sense to assign the power of fecundation to the liquid part of the semen or to some principle given off by the semen. Girolamo Fabrici d'Aquapendente (ca. 1533–1619) had spoken of a seminal spirit in 1621, and Swammerdam first used the term *aura seminalis* in 1685. Until the semen could actually be observed in the uterus, it made more sense to assume that the fertilization was accomplished by a spirit or influence.

One group of microscopists, the "animalculists," beginning with Leeuwenhoek and Nicolaas Hartsoeker (1656–1725), believed that the sperm did indeed reach the uterus and that they contained the preformed embryos. Hartsoeker even described the "homunculus," or preformed embryo, that he believed had to exist in the head of the human sperm. Since the animalcules were present in great numbers in the semen, very few of the preformed embryos they contained would ever grow to birth. There had to be a great destruction of potential animals. It did not seem that God would be so wasteful of his creatures. It was also hard to explain how the mother could pass on her characteristics to an embryo coming entirely from the male, and therefore the animalculist view declined in the early eighteenth century. Some even denied the existence of the animalcules. Linnaeus (a great naturalist, but a poor microscopist) said that they were inert masses of fatty material; others denied that they swam on their own or else claimed that their only purpose was to stir the semen. Bonnet's discovery of parthenogenesis contradicted the animalculist theory completely. Even Hartsoeker, the most outspoken of the animalculists, renounced the theory of preformation in 1722 after he had performed experiments on regeneration. If a crayfish could easily replace an amputated leg or claw, then the same intelligence responsible for regenerating the claw could also form an entire animal. Regeneration made any simple preformation theory untenable.

By 1744, when Trembley did his experiments on the polyp, theories of generation were becoming more sophisticated. The preformed human embryo, if it existed, was no longer thought to be a miniature human curled up in part of the ovaries or in the body of the animalcule. It merely had to carry in some fashion the form or plan from which the embryo could be built. It did not have to resemble the adult any more than a blueprint resembled a house. Considered from this point of view, the theory of preformation was not far from the truth.

The theory of epigenesis was equally sophisticated. It no longer required Aristotle's concepts of substance and form. It could be understood as a statement about proper scientific method. The chick gradually appeared in the undifferentiated yolk of the fertilized egg. No form could be observed in the yolk of a freshly laid egg. To claim that it contained a preformed embryo was to claim what could not be seen. What could be seen were the organs of the chick gradually appearing as the yolk incubated. From this point of view epigenesis was a statement about the need for caution in drawing conclusions from experiments in embryology. It refused to imagine structures that could not be seen or otherwise demonstrated.

Neither position can be taken as absurd; each has its own compelling logic. From our modern perspective we cannot say that one theory was right and the other wrong. In the eighteenth century embryologists had to work their way through this maze to create a theory that would not offend either their logic or their senses.

Philosophers of the eighteenth century attempted to understand the origins of living things by studying the phenomena of reproduction and regeneration. An alternative would have been to study heredity, the laws by which characteristics appear in subsequent generations. It was a subject that could have been investigated experimentally with the knowledge and equipment available in the eighteenth century, but because the mechanism of generation was such an important philosophical problem, the possibility of reducing the transmission of characteristics to rule was largely overlooked. Stock breeders built a fund of information about heredity, but they were more concerned about practical results than scientific principles. Also, the large animals that they raised were not the most suitable subjects for studying heredity; they bred slowly and were expensive to maintain. Plants, once their sexuality had been recognized, made good subjects because they could be bred much more rapidly and cheaply.

The first studies of heredity in plants were investigations of hybrids. Linnaeus searched for hybrids and thought that he had found many, but he never tested them for purity of type. He classified any plant intermediate between known species as a new species without checking to see if it bred true. Much more thorough experiments (on tobacco plants) were performed by Joseph Gottlieb Kölreuter (1733–1806) beginning in 1760; his results were published as *Vorläufige Nachricht von einigen das Geschlecht der Pflanzen betreffenden Versuchen und Beobachtungen* [*Preliminary report on experiments and observations of certain species of plants*] in 1761, with later supplements. Kölreuter carried out more than five hundred different hybridizations and described the pollen of more than a thousand plant species. Because he believed that the order of nature required the fixity of species, he performed his experiments in order to discover why the existing order of nature was not swamped by innumerable new hybrid species.

He found the answer when his tobacco hybrids proved to be sterile; all of the flowers fell off, and the plants produced no seed. It was, for him, "one of the most wonderful of all events that have ever occurred upon the wide field of nature." Kölreuter's hybrid was, as he said "the first botanical mule which has been produced by art."[4] By backcrossing his hybrids with the original nonhybrid species, he was able to create second-generation hybrids. He later discovered that first-generation hybrids of some plants such as pinks, carnations, and sweet williams were partially fertile and therefore was able to obtain true second-generation hybrids by self-fertilization of his first-generation hybrids. The results were puzzling. The first-generation hybrids were always the same, whereas the second-generation and backcrossed hybrids showed a bewildering variety.

Kölreuter concluded that the lack of orderly characteristics in second-generation hybrids was the result of man's interference with nature. By bringing together plants from different parts of the world and by pollenating them artificially, the naturalist was creating conditions that would never occur in nature. Because the hybrids did not breed true, he concluded that they could not create new species. To explain his results, Kölreuter imagined a chemical analogy. Male and female "seed materials" united in the plant in a way similar to the union of an acid and an alkali to form a crystalline salt.

The variety of second-generation hybrids was a result of the great variety of proportions in which the seed materials could combine.

A similar conclusion was reached by Pierre Maupertuis, the same Maupertuis who invented the principle of least action and led the expedition to Lapland to study the curvature of the earth (see Chapter II). Whereas Kölreuter studied heredity in plants, Maupertuis worked with animals. His interest in heredity was stimulated in 1743 when an albino Negro was exhibited in Paris. There was a great interest at the time in physical abnormalities, both as carnival attractions and as clues to the mechanism of generation. Although Maupertuis's previous work had been almost entirely mathematical, he had always kept a large number of pets and bred them to obtain special characteristics. He concluded, as did Kölreuter, that both parents contributed to the formation of the offspring and that the embryo could not be preformed in either parent. Instead he suggested that both the male and the female produced a semen containing special particles that mingled to create the embryo. Like Kölreuter, he believed that fertilization was a dynamic chemical process similar to the formation of an elaborate crystal.

Because the offspring could resemble either parent or even a grandparent in any part, the particles in the semen must be collected from all parts of the parents' bodies and even be carried over from previous generations. Maupertuis suggested that if the genetic particles had to migrate from all parts of the body, a mutilation of one part consistently, through several generations, would probably cause the defect to become heritable. Some abnormalities were extreme enough to make survival of the individual impossible, but if the abnormalities were small and if these individuals were to mate with one another rather than with normal individuals, then their abnormal characteristics might become permanent in their posterity. In this way an entire race of albino Negroes could appear.

In order to learn more about the transmission of abnormal characteristics, Maupertuis investigated polydactylism in humans, which is the appearance of extra digits on either or both hands or feet. Réaumur had suggested breeding fowl with different numbers of digits to discover how this characteristic was transmitted, but Maupertuis said that the information could be obtained more easily from human families that were polydactylous. In his *Lettres de M. Maupertuis* [*Letters from Monsieur de Maupertuis*] (1751) and his *Système de la nature* [*System of nature*] (1757), Maupertuis gave the genealogy of the Ruhe family of Berlin. The data that he obtained, covering four generations, showed that the trait could be transmitted by both the male and the female and that its occurrence in four consecutive generations could not have been accidental, since the probability of its occurring accidentally was astronomically small. Réaumur obtained similar information on the Kelleia family, also polydactylous, with similar results.

Maupertuis's contemporaries mention other breeding experiments that he performed, one in particular in which he duplicated an unusual marking on an Icelandic dog by breeding its puppies for several generations. If Maupertuis had chosen a more suitable subject, had studied more cases, and had measured the frequency of the characteristics among his samples rather than the probability of their appearance among the population as a whole, then he might have obtained valuable rules for the transmission of characteristics, but he did not seek that kind of information. His interest was in embryology, not in heredity as such, and once he had demonstrated that abnormal characteristics such as polydactylism must indeed be inherited, he made no effort to reduce the frequency of their appearance to any general rule.

Moreover, Maupertuis's experiments did not investigate the order of nature but deviations from that order. It was not obvious in the eighteenth century that deviations from nature could be or should be reduced to rule. Kölreuter, we noted, was greatly relieved to discover that hybrids did not disrupt the order of species. The "Detailed explanation of the system of human knowledge" in the *Encyclopédie* divided all of natural history into the "order of nature" and "deviations from nature." These were two separate categories, one set of observations evidencing order, the other evidencing disorder. Maupertuis was trying to learn about the mechanism of generation by studying abnormal cases. If he had hoped to discover rules of heredity, he would have studied normal cases.

Buffon adopted a theory of generation similar to that of Maupertuis. In the second volume of his *Histoire naturelle* [*Natural history*] (1749), he brought forward his theory of organic molecules, interior mold, and penetrating force. He also adopted the theory of the double semen. He believed that there was no female ovum but that a semen produced by the Graafian follicles in the ovaries mixed with the male semen in the uterus to form the embryo. The vital part of both semens was composed of organic particles that were in excess of the needs of the interior mold and were stored in the ovaries and testicles for the future production of offspring. Buffon believed that the reason why individuals ceased to grow at puberty was that the body at that time no longer needed the organic molecules and they could begin to accumulate in the reproductive organs.

Because the organic molecules were distributed throughout nature and were never themselves destroyed, they could appear in a variety of microscopic forms. Buffon claimed that the small animals observed in putrefying broth were groups of organic molecules released from dead material by the absence of an interior mold. Buffon persuaded John Turberville Needham (1713–81), an English microscopist, to perform experiments on spontaneous generation in beef broth and hay infusions. Needham believed that the enormous number of little animals he observed could not all have come from seeds in the infusions. To see if they came from the outside or were generated in the liquid, he heated his flasks of broth to kill all the animals and corked the flasks tightly. He found that after a short time there were again numerous objects swimming in the broth. Buffon concluded that the objects were not real animals at all but collections of organic molecules in various degrees of organization. The experiments were difficult, and Needham had been careless. The infusoria either had not been killed by the heat, or the corks that he used to seal the flasks were not tight.

Lazzaro Spallanzani carried out experiments in 1765 and again in 1776 to check Needham's results. He used flasks with slender necks that could be melted shut, guaranteeing a seal against organisms entering from the outside. Spallanzani found that boiled broth in a sealed container would remain sterile indefinitely but that if he broke the neck of the flask animals soon appeared in the liquid in great numbers. He attacked Buffon's assumption that the objects observed in the infusion were not real animals. Buffon had said that these infusoria were only the remains of animals and observed that they had lost their tails and did not move under their own power but were moved about mechanically by the liquid.

Spallanzani showed that the tails were still there but merely rolled up in the cases that Buffon had observed. He saw the little animals move, navigate, ingest food with cilia, and reproduce in a variety of ways. In short, they were true animals, not just collections of organic molecules.

The Revival of Preformation

By the time Spallanzani did these experiments, the preformation theory was staging a comeback. Just before its revival, however, one of the strongest appeals in favor of epigenesis was made by

Caspar Wolff (1734–94), a follower of the chemist Stahl and the Leibnizian philosopher Christian Wolff. In his *Theoria generationis* [*Theory of generation*] (1759) and later in his "De formatione intestinarum" ["On the formation of the intestine"] (1768–9), Caspar Wolff argued that the embryo was created by a *vis essentialis* (essential force) inherent in living matter. In the tradition of Leibniz's dynamic philosophy, he held that matter was essentially active, but he made no claim to a detailed knowledge of this force. He wrote: "We may conclude that the organs of the body have not always existed, but have been formed successively – no matter how this formation has been brought about. I do not say that it has been brought about by a fortuitous combination of particles, a kind of fermentation, through mechanical causes, through the activity of the soul, but only that it has been brought about."[5] This positivist attitude was characteristic of the new theories of generation.

Albrecht von Haller had begun as an epigenesist, but after carrying out his own experiments on the chick embryo he was converted to preformation. With his knowledge of physiology, Haller recognized the extent to which the organs were all interdependent, and he could not believe that they appeared successively in the embryo. A heart or liver by itself could not live and function. Therefore the organs had to appear together, even if they were observed successively. By using different reagents to harden the parts of specimen embryos to produce greater contrast, Haller showed that the developing embryo had greater differentiation in the early stages than one would conclude from simple observation. If the organs had appeared all together, one would have had to conclude that they existed in some preformed state.

Beginning in 1760, the three best experimentalists of the century – Bonnet, Haller, and Spallanzani – were all drawn to the ovist version of the preformation theory. Bonnet was the most speculative of the three. His eyesight had failed after his important work on parthenogenesis in the 1740s, and he had turned his efforts to finding a logical mechanism of generation. In his *Contemplation de la nature* [*Contemplation of nature*] (1764), Bonnet defined preformation in a way that shows the flexibility and abstractness of the theory: "I understand by the word 'germ' every pre-ordination, every preformation of parts capable by itself of determining the existence of a plant or animal." Regeneration produced problems for the preformation theory, but Bonnet urged that the word *germ* "be taken in its widest sense." Even the polyp could be said to regenerate from a germ if the germ were defined as any "secret preorganization." Moreover, the germ in the female did not fully

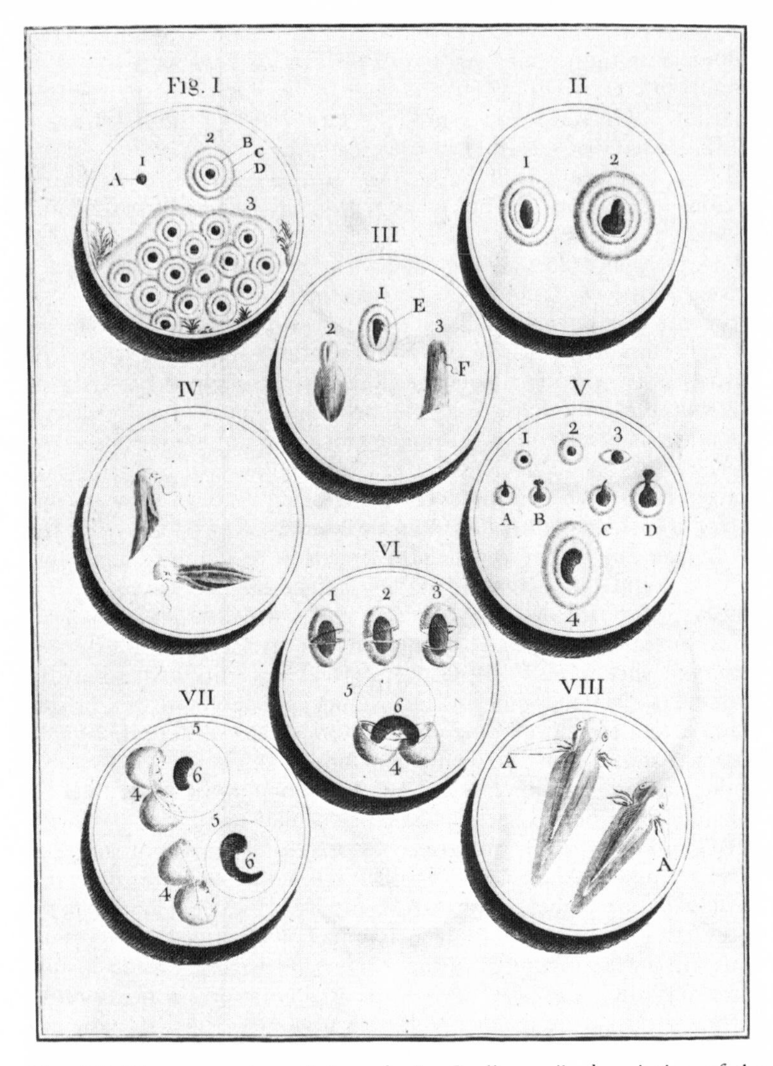

Fig. 5.5. The generation of the tadpole. Spallanzani's description of the development of the tadpole of the green frog. Figure I shows a mass of eggs, one of which is magnified to show its structure. The other illustrations show different stages of development. In these illustrations Spallanzani wishes to show that the tadpoles do not come from true eggs but are preformed in the mother's body. He writes that "as greater deference was due to what nature shewed me so plainly . . . than to the authority of the most celebrated writers, I am obliged to call these globules tadpoles or

determine the embryo; the action of the semen and the process of nutrition created the variations observed in every species. A germ, for Bonnet, was "a miniature man, a horse, a bull etc., but it [was] not a *certain* man, a *certain* horse, a *certain* bull, etc."[6] The individual variations came from outside the germ.

The best experiments on generation during the entire eighteenth century were those done by Spallanzani. After revealing the errors in the observations of Buffon and Needham, Spallanzani extended his experiments to spermatozoa. Buffon had argued that the spermatozoa were like the infusorians that Needham had observed in broth; they were merely clumps of organic particles caused by decay, not real animals. Experimenting with canine semen, Spallanzani showed that the spermatozoa were in the semen from the time it was taken from the dog and that Buffon's observations on semen were extremely misleading. Because Buffon had not used fresh semen, he had not been observing the spermatozoa at all but had described infusorians in the putrefying semen. Spallanzani was even able to observe the sperm in place in the transparent vas deferens of a live fasting salamander.

His most important experiments investigating the nature of sperm were performed on frogs. Frogs were especially good subjects for these experiments because they fertilize their eggs externally. As the female releases the eggs in water, the male sprays semen on them. Eggs touched by the semen produce tadpoles, whereas those taken from inside the body of the female are sterile. Spallanzani made tight-fitting taffeta pants for the male frogs to contain the semen. The frogs were then allowed to mate normally. None of the eggs developed. The story of Spallanzani's taffeta frog pants often elicits laughter, but it was an important and difficult experiment. Réaumur and Nollet, both able experimenters, had tried to carry out this same experiment and had failed. Once he had demonstrated that the fertilization of frogs' eggs was external, Spallanzani was able to use artificial insemination, which gave him much greater experimental control.

He took semen from the seminal vesicles of male frogs and painted it onto unfertilized eggs taken from the female. The unpainted eggs

---

Caption to Fig. 5.5 *(cont.)*
fetuses instead of eggs; for it is improper to name any body an egg which, however closely it may resemble one, takes the shape of an animal without leaving any shell." *Sources:* Lazzaro Spallanzani, *Dissertation relative to the natural history of animals and vegetables,* 2 vols. (London, 1784), vol. II, pl. I. By permission of the Syndics of Cambridge University Library.

decayed, but the painted eggs produced tadpoles. As he improved his experimental technique he found it more convenient to place the semen on the eggs with a needle, scratching the outer surface – a change that produced unexpected confusion in his results, as we shall see.

Spallanzani next set out to determine what part of the semen was responsible for fertilizing the eggs. His first experiment disposed of the *aura seminalis*. He attached twenty-six eggs to a small watch glass, which he suspended over another watch glass containing fresh semen. The eggs were placed as close as possible to the semen without actually touching it and were obviously being bathed in any aura that might be leaving the semen. The eggs remained sterile. He then determined the external circumstances that might affect the sperm. Vacuum, cold, and oil had no effect. Heat, evaporation, wine, and filtering destroyed its fertilizing ability. In 1784 he published the results of experiments in which he had filtered the semen. The liquid portion of the semen would not fertilize eggs; the thick portion containing the sperm did fertilize eggs. From our perspective, this experiment should have been conclusive; we would conclude that Spallanzani had shown that the "spermatic worms" were the actual agents of fertilization – but this is not what Spallanzani concluded. He believed that the liquid left on the filter paper was responsible for fertilization and that the spermatic worms were just that – parasites in the semen.

This case has often been put forward as an example of an experimenter blinded by a previous commitment to an erroneous theory – in this case preformation. But Spallanzani had reason for caution. In the first place, he performed the experiments not to separate out the sperm but to separate two fluids in the semen, one that he found in the seminal vesicles and called the "seed," the other, a more dense liquid, that he called the "juice of the testicles." In the second place, the sperm did not appear to be responsible for fertilization. On two occasions he had placed sperm-free liquid from the semen on eggs, and they had developed. On another occasion semen treated with wine to kill the sperm had also fertilized eggs. It is possible that Spallanzani's supposedly sperm-free liquid actually contained sperm, but this is unlikely considering his skillful experimental technique. In 1910, Jean-Eugène Bataillon (1864–1953) showed that frogs' eggs could be made to develop parthenogenetically by pricking them with a glass needle or micropipette. Probably Spallanzani's technique of applying semen to the eggs with a needle sometimes caused parthenogenesis of unfertilized eggs.

The decline of the preformation theory during the first half of the eighteenth century and its revival in the 1760s tells much about what was happening in the life sciences. In its earliest form, preformation was the only answer that the mechanical philosophy could give to the problem of generation. The preformed embryo was regarded as a completely formed animal that needed only to grow to become an adult. The theory explained generation without resorting to special vital forces. But the mechanical philosophy proved unable to explain vital phenomena. The emphasis on structure gave way to an emphasis on vital function and to a phenomenalist experimental approach.

The preformation theory that reappeared in the 1760s was quite different from that of the seventeenth century. By this time the idea of the preformed embryo had become an abstract concept applying to any preexisting order, form, or mold that gave form to the embryo. Moreover, the theory implied that living things were different from nonliving things and no longer constituted a strictly physical-chemical explanation of life.

Natural History

For natural historians to make any sense out of the multitude of natural forms, they must first reduce these forms to some kind of order or classification, and that classification will be arbitrary to a certain extent. One could, for instance, choose to list all plants by classifying them in terms of an essential characteristic such as the flowering parts. This might help distinguish among different forms, but it would not describe any form in its entirety. Distinguishing between plant forms on the basis of a single characteristic would therefore be "artificial." The goal of naturalists in the eighteenth century was to find a "natural" system, one that identified plants and animals by their "essences" – that is, those things that made them what they were. The essence of man was his rational soul, not the color of his eyes. According to Aristotle, the first was an essential property, the second a mere accident. In Christian terms, the search for a natural system was a search for God's plan. No one doubted that the forms of living things were related in some harmonious way to fulfilling God's purposes in his creation.

There was great difficulty, however, in deciding what constituted the essence of a plant or animal, and consequently which systems of classification were natural. The immediate problem was whether a natural system required a single characteristic or a whole complex of characteristics to define a species. A single characteristic such as the leaf or stem might serve to distinguish among forms in one part of the plant kingdom and fail completely in another part. An alternative was to subordinate certain characteristics to others. The shape of the flower might be made the principal differentiating characteristic; other characteristics would be made subordinate to this major one and used for further differentiation among plants that had similar flowers. Deciding which characteristics were dominant and which were subordinate, and in what order, inevitably involved a certain amount of arbitrariness, which raised doubts about whether the system was natural.

Aristotle's formal logic was based on a systematic arrangement of things into categories and classes, and one would therefore have expected Aristotle to apply it in his natural history. But Aristotle was also an acute observer, and he soon realized that the principal divisions required by his logic did not apply to living things. He concluded that the entire complex of characteristics defined the species and that therefore the entire complex had to enter into the system of classification.

Aristotle's influence still dominated natural history in the seventeenth century, but that domination was broken in taxonomy by the discovery of the sexuality of plants. Since most plants are hermaphrodites, containing both male and female reproductive organs, the fact that they reproduce sexually was not obvious, and only a few cases, such as that of the date palm, were known to antiquity. Rudolph Jacob Camerarius (1665–1721) was the first to demonstrate experimentally the sexuality of plants. In his *Epistola . . . de sexu plantarum* [*Letter on the sex of plants*] (1694), he showed that in order for plants to bear fruit the pistils of the female flowers had to be provided with pollen.

The sexuality of plants provided a possible basis for a natural system of classification because the mechanism of generation must of necessity also determine the plant's form. Camerarius was not himself a taxonomist or even an especially prominent naturalist, but his discovery gave support to the earlier very important system of Andrea Cesalpino (1519–1603), which was based on the plant parts involved in fructification, and to the more recent system of Joseph Tournefort, who chose the reproductive organs of plants – that is their flowers and fruits – as the only reliable characteristics that could form the basis for classification. In the case of animals, it was more obvious that species were determined by their ability to reproduce, and therefore the analogy to plants lent further support to the notion that any natural system should be based on the species's reproductive characteristics.

John Ray struggled with the problem of the natural system in his

Methodus plantarum nova [*New method of plants*] (1682), his *Historia plantarum* [*History of plants*] (1686), and his later *Methodus plantarum emendata* [*The Method of plants emended*] (1703). For simplicity of classification he originally followed Cesalpino and the systematists, but the methods of British natural philosophy at the time persuaded him that no single characteristic could create a natural system. These different points of view became explicit in a controversy between Ray and Tournefort that we will examine shortly, so that during the Enlightenment there existed two sharply divided camps – those who believed in the possibility of a natural system based on a single characteristic and those who insisted that a complex set of characteristics was necessary. A typical statement of support for the latter type of classification is this passage from Michel Adanson's (1727–1806) *Familles des plantes* [*Families of plants*] (1763–4): "The botanical classifications which only consider one part or a small number of the parts of the plant are arbitrary, hypothetical and abstract, and cannot be natural . . . Without doubt, the natural method in Botany can only be attained by consideration of the collection of all the plant structure."[7] Linnaeus, on the other hand, argued that "systematic division of the plants should take as its basis the primary structure. Therefore, as Nature confirms that fructification is the only systematic foundation of Botany, it can thus be demonstrated to be the absolute foundation. This has been accepted by the greatest systematists as the prop and mainstay of Botany."[8]

Linnaeus was undoubtedly the greatest botanist of the eighteenth century, and probably of all time. He obtained his botanical training in Sweden, traveled to Holland to obtain a medical degree, and subsequently went to Leiden, where he worked with Boerhaave. It was in Leiden that his famous *Systema naturae* [*System of nature*] appeared in 1735. In that work and in two subsequent works – *Fundamenta botanica* [*Foundation of botany*] (1736) and *Classes plantorum* [*Classes of plants*] (1738), Linnaeus used the characteristics of fructification to classify plants in a system that was more precise and useful than any previously devised (see Figure 5.6). He recognized that his system was not completely natural and constantly attempted to improve it, but he did not doubt that some system was necessary for botany and that a natural system did indeed exist. His classification began with the species, which he believed to have been fixed from the time of Creation. But even here his own experiments on hybridization raised doubts, and in the last edition of the *Systema naturae* he no longer insisted on fixed species.

Another of Linnaeus's contributions was the binomial nomencla-

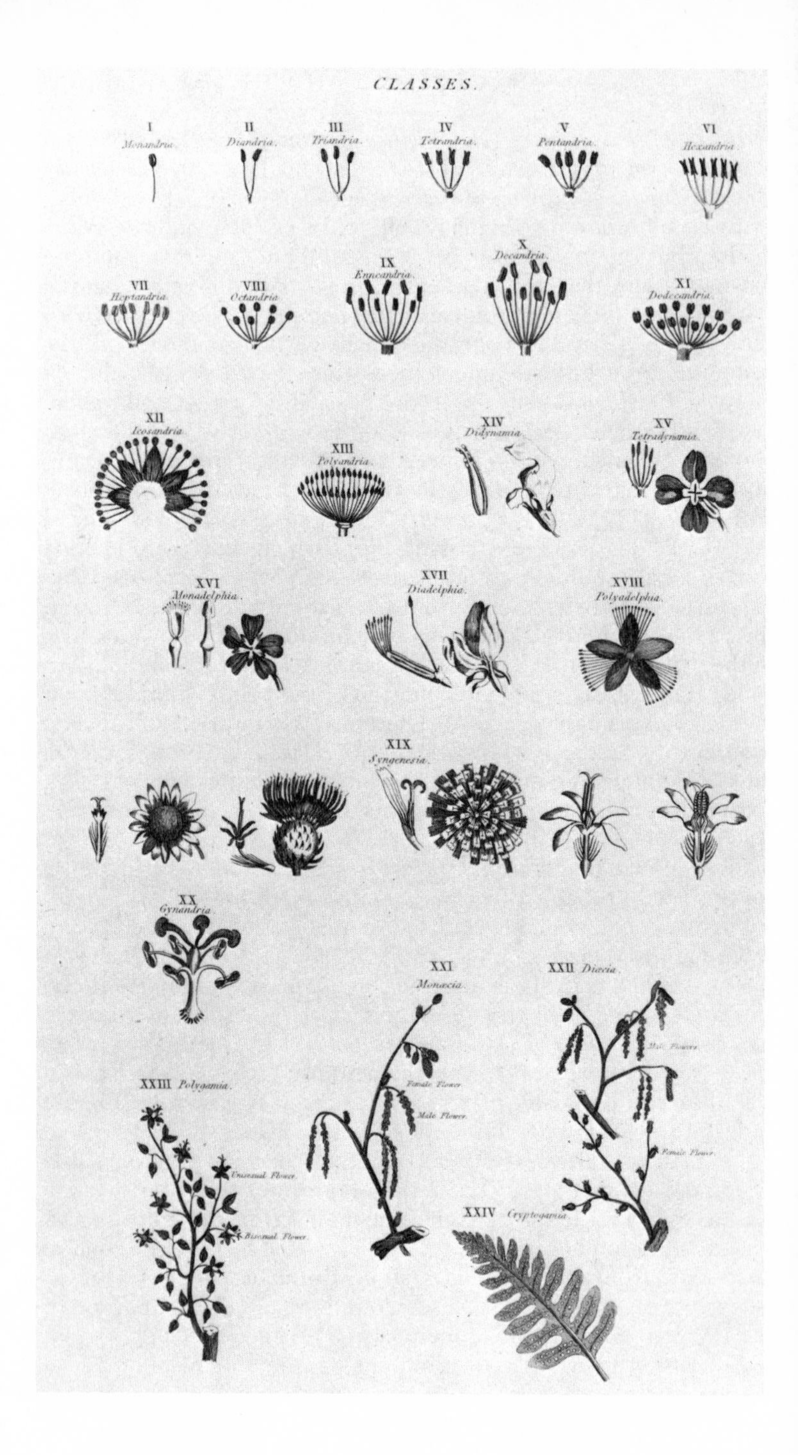
CLASSES.
I Monandria.
II Diandria.
III Triandria.
IV Tetrandria.
V Pentandria.
VI Hexandria.
VII Heptandria.
VIII Octandria.
IX Enneandria.
X Decandria.
XI Dodecandria.
XII Icosandria.
XIII Polyandria.
XIV Didynamia.
XV Tetradynamia.
XVI Monadelphia.
XVII Diadelphia.
XVIII Polyadelphia.
XIX Syngenesia.
XX Gynandria.
XXI Monœcia.
Female Flower.
Male Flower.
XXII Diœcia.
Female Flower.
XXIII Polygamia.
Unisexual Flower.
Bisexual Flower.
XXIV Cryptogamia.

ture that he introduced into botany in 1753 and later into his classification of animals. The first Latin word in the name identified the genus and the second Latin word identified the species, a reform of taxonomic language comparable to the new chemical nomenclature of the French chemists at the end of the century, and just as permanent.

The debate between the supporters of a taxonomy based on a single characteristic and the supporters of a taxonomy based on a complete set of characteristics reached its climax in the first volume of Buffon's *Histoire naturelle.* Buffon began his attack on the Linnaean system at the Paris Academy of Sciences in 1744, just as that system was obtaining almost universal acceptance from botanists. Moreover, Buffon criticized not only Linnaeus's system but all systems of classification that depended only on external characteristics. He believed that the universe was made up of individual objects. To force them into a rational set of categories was to impose an artificial abstraction of the human mind on nature. He wrote: "The more one increases the number of divisions in natural things, the closer one will approach the truth, since there actually exist in nature only individuals. . . . The Genera, Orders, and Classes exist only in our imagination."[9]

Buffon turned from systematic taxonomy to the image of the Great Chain of Being, a view of nature that had originated with Aristotle and had been employed by Leibniz in his metaphysical system. The Great Chain of Being, or Scale of Nature, was a linear, hierarchical progression of forms stretching from the simplest to the most complex. Leibnizian metaphysics demanded that it be continuous and full. There could be no gaps in the chain and no marked transitions between forms. Buffon described it as a chain

---

Fig. 5.6. The classification of plants according to Linnaeus. Linnaeus divided plants into twenty-four classes according to the character of their flowering parts. The number of stamens determines the first eleven classes, and the shape of the stamens determines the next nine classes. Plants in the next three classes have stamens and pistils in separate flowers. The twenty-fourth class consists of plants that lack true flowers. This is an "artificial" system because it employs only a single characteristic (the stamens and pistils of the flowers). Even though that one characteristic may be "essential" – that is, necessary for the plant to be what it is – the system cannot be "natural," because it ignores the multiplicity of characteristics that determine the plant form. *Sources:* Robert John Thornton, *A new illustration of the sexual system of Linnaeus* (1799–1807). By permission of the Syndics of Cambridge University Library.

of degradation descending from man at the top: "One can descend by almost insensible degrees from the most perfect creature [man] to the most disorganized matter. . . . It will be seen that these imperceptible gradations are the great work of Nature; one will find such gradations not only in size and form, but also in the motions, the generation, and the succession of each species . . . It will clearly be perceived that it is impossible to give a general classification, a perfect systematic arrangement, not only for Natural History as a whole, but even for a single one of its branches."[10] At first glance this criticism seems misguided (How can a naturalist work without any classification?) and incomprehensible in light of Buffon's past career as a mathematician and Newtonian physicist. In the *Premier discours* [*First discourse*] to his *Histoire naturelle,* he not only condemned the Linnean system but also condemned mathematics as an abstract creation of the mind not corresponding to nature as it really exists.

Buffon's criticism makes more sense if we see it as a continuation of the debate over taxonomy that began at the beginning of the century between Ray and Tournefort. Ray had directly revealed the source of his doubts – John Locke's *Essay concerning human understanding* (1690). Buffon was less explicit a half century later, but his contemporaries recognized nevertheless that standing behind Buffon's "untenable pyrrhonism" were "the doctrines of Mr. Locke."[11]

In his later *Methodus plantarum emendata* (1703) Ray had argued the Lockean position that the essences of things are wholly unknown to us and that we obtain knowledge of nature only through our senses. Thus we receive only collections of sensations, none of which can be the essence of the object we perceive. Reflecting on this multitude of sensations, we make judgments about essences. Just as the secondary qualities of taste, smell, color, and so forth are in our way of perceiving objects, not in the objects themselves, so the external characteristics of plants are mere indications. They cannot be the essences themselves, and therefore no single characteristic can be the basis of a natural system of classification. By considering the entire complex of characteristics, we can make the best judgment about the relationships among different plant forms, but even if we use all of the characteristics, our knowledge of essences can only be probable knowledge, which can never reach the certainty of mathematics. When Buffon criticized Linnaeus's taxonomy, in 1749, he reechoed the arguments of Locke and Ray. Linnaeus's taxonomy, Buffon claimed, shared the weakness of mathematics. It was abstract, artificial, and precise, because it came from the mind, not from nature. It obtained precision at the expense of realism.

Buffon's answer was to determine species not by any characteristic but by their reproductive history. He adopted the reproductive characteristic used by Ray and Réaumur. Two individual animals or plants are of the same species if they can produce fertile offspring. The members of any single species can, of course, be identified by some physical characteristics, but those characteristics can only be accidental properties. The essential identification of the species is the history of its propagation, not any physical form. Thus, according to Buffon, "species is an abstract and general term . . . to which a corresponding object exists only in considering Nature in the succession of time, and in the constant destruction and renewal of beings."[12] The meaning of *history,* in the term *natural history,* has here taken on a temporal dimension. Buffon argues that we know the essence of natural things only through their succession in time. If we know species only by the history of their propagation, then it is absurd to use the same principles for classifying living and nonliving things. Rocks do not mate and have offspring. The taxonomy of the mineral kingdom cannot be based on the same principles as that of the animal and vegetable kingdoms.

Nevertheless, Buffon did not intend to limit history to plants and animals. His natural history was both temporal and cosmic. It was not limited to a pure description of things as was, for example, Réaumur's marvelously detailed *Mèmoires pour servir à l'histoire des insects* [*Memoirs on the history of insects*] (6 vols., 1734–42) but revealed the whole scheme of nature. Thus the first volume of Buffon's *Histoire naturelle* began with a history of the earth.

Buffon borrowed his inspiration for a temporal history from the physico-theologians of the beginning of the century. For them also natural history was temporal, because they saw God working his Providence through time. But for Buffon, natural history was entirely natural. His history of the earth simply ignored Genesis and biblical chronology. Throughout his life he altered his estimates of the age of the earth, but they were always much greater than the six thousand years calculated from the biblical story. In his *Epoques de la nature* [*Epochs of nature*] (1778), he divided the earth's history into seven epochs. The earth was originally a molten mass torn away from the sun by a colliding comet about eighty thousand years ago. Because of its smaller size, the earth cooled faster than the sun. Its surface solidified as gases were vented into the atmosphere. Then, as it cooled, the solid crust shrank and cracked, creating the oldest valleys and mountains. When the temperature dropped far enough for the vapors to condense, seas formed, eroding the mountains and depositing sediments. The first forms of life appeared, leaving their fossil remains in these sediments. The seas then retreated, since much of the water disappeared through rents in the cooling crust. Plants appeared on land, while volcanoes, fueled by the organic matter that was washed into crevices in the earth, changed the landscape. Land animals appeared as the earth continued to cool. The continents separated, and new islands arose in the Atlantic. Finally, in the seventh epoch, man appeared and began to control and shape the earth. Within another ninety-three thousand years the earth will become too cold to support life. The key to all of these changes was the cooling of the earth. Buffon heated globes of cast iron and measured their rates of cooling. From this information he extrapolated to a globe the size of the earth and estimated the time for each epoch. Today his history appears fanciful in the extreme, but it employed natural causes and gave a temporal dimension to natural history.

The historical dimension in Buffon's writing separated him from his contemporaries. Charles Bonnet, for example, also believed in the Great Chain of Being, but it was a chain without a temporal history. In his *Considérations sur les corps organisés* [*Considerations on organized bodies*] (1761) and his *Contemplation di la nature* [*Contemplation of Nature*] (1764), he reiterated the principle of plenitude. There were no gaps or demarcations between the forms of living things; classifications were entirely nominal. For Bonnet the continuity of the chain was necessary to guarantee its rationality. Leibniz had made the same argument fifty years earlier, and for Leibniz it had a mathematical foundation. Leibniz believed that the forms of things stood in relationships similar to those that exist in mathematics. If the forms were discontinuous, then the functions relating them would be discontinuous, and in mathematics at mid-century a discontinuous function had no meaning at the point of discontinuity. Rationality required continuity.

Bonnet's rational philosophy had no place for time. Like logic and mathematics, it stood outside of time and was not contingent on temporal events. Bonnet included minerals in the Chain of Being because they were merely the simplest forms in the chain. His belief in the preformation of germs was also consistent with his view of nature as static. Buffon, on the other hand, insisted on the temporal dimension, criticized the use of mathematical analogies in natural history, refused to include minerals in the Chain of Being, denied preexistent germs, and included historical geology. Buffon changed the meanings of both the terms *nature* and *history* in natural history.

Buffon's separation of the study of the earth from natural theology was characteristic of geologic method during the second half of the eighteenth century. In Fact the terms *geology* and *geologist* were first used regularly with their modern meanings by Horace Bénédict de Saussure (1740–99), author of *Voyages dans les Alpes* [*Voyages in the alps*] (1779–96), one of the earliest geologic field studies of the Swiss Alps (see Figure 5.7). Jean Guettard, who, as we saw earlier, employed Lavoisier in his geologic survey, was the first to realize the extent of volcanic geology in Europe. He recognized that the black milestones that he encountered near Moulins in central France were probably volcanic in origin, and he traced them back to the quarry from which they came. He identified the quarry as an old lava flow and determined that the abrupt rocky mountains in central France called *puys* were cores of old volcanoes. Nicolas Desmarest (1725–1815) continued to seek evidence of volcanic activity in Europe and showed that it was much more extensive than even Guettard had supposed. More important, Desmarest concluded from studying different kinds of rock associated with volcanoes that basalt was of igneous origin.

The emergence of geology as a science at the end of the eighteenth century is usually associated with a German teacher of mining, Abraham Gottlob Werner (1749–1817), and with the Scottish philosopher James Hutton (1726–97). The views of these men are usually described as representing another dichotomy in the history of science, like that between mechanism and vitalism, or between the theory of preformation and the theory of epigenesis – the assumption being that one side was right, the other wrong, and that right inevitably triumphed over wrong. In fact, their views were not entirely opposed.

Werner believed that rock strata were either sediments originally deposited at the bottom of the sea or crystalized deposits precipitated from seawater. Werner gave only a minor role to volcanic action. This emphasis on water as the chief agent of rock formation caused Werner and his colleagues to be named "Neptunists." Those who emphasized volcanic action were called "Vulcanists" or "Plutonists."

Hutton believed that the warping and tilting of strata was caused by the earth's internal heat, which also was vented occasionally through volcanoes. More important, he believed that basalt had crystalized as it cooled from a molten state and was not, as Werner had thought, formed by precipitation from the seas, although Hutton readily admitted the role of water in eroding the land and in depositing sediments. Hutton was the first to state clearly, in his *Theory of the earth: with proofs and illustrations* (1795), that the earth

Fig. 5.7. The geology of the Alps. The first description of geologic fieldwork in the Alps was Horace-Bénédict de Saussure's *Voyages dans les Alpes* [*Voyages in the Alps*] (1779–96). In this illustration from the book, the two climbers are dwarfed by the immensity of Mont Blanc. De Saussure concluded that the distorted strata seen in the mountains could only have been created by explosive forces deep within the earth. *Sources:* Horace-Bénédict de Saussure, *Voyages dans les Alpes précédés d'un essai sur l'histoire naturelle des environs de Genève* (Neuchâtel, 1779–96), vol. IV, pl. IV. By permission of the Syndics of Cambridge University Library.

changed only slowly and uniformly by the processes that have been observed during historic time. "We find," he wrote, "no vestige of a beginning – no prospect of an end." This "uniformitarian principle" can be juxtaposed against a doctrine of "catastrophism" – the assumption that land forms were caused by geologic events greater than any observed by man.

The Noachian Flood was one such catastrophic event, and a "Neptunian" one to boot. Catastrophes also helped to fit geologic time into the chronology given in the Bible for human history. But the developments of geology in the nineteenth century demanded a geologic time scale much greater than any that the Bible would allow. It also became clear that basalt had not precipitated from the seas but had an igneous origin. Both of these developments supported the Vulcanist and the uniformitarian positions, and this supposed victory of the Vulcanist-uniformitarians over the Neptunist-catastrophists has been seen by some as a triumph of science over religion and as another instance of the superiority of British science. In fact, Hutton was profoundly religious, although not an orthodox Christian. In keeping with the natural theology tradition, he believed that the erosion cycle was God's means of replenishing the soil and providing mankind with food. Nor was he totally averse to catastrophes. He stated in his *Theory of the earth:* "The theory of the earth that I here illustrate is founded on the greatest catastrophes which can happen to the earth, that is [continents] being raised from the bottom of the sea and sunk again."[13] John Playfair (1748–1819), who popularized Hutton's theory in his *Illustrations of the Huttonian theory of the earth* (1802), removed both the catastrophes and the action of God from the theory. Since the supreme being was no longer allowed to move the continents up and down, this particular part of Hutton's theory remained unexplained.

A less artificial distinction between the works of Werner and Hutton than that which implies that they were totally opposed in their beliefs can be found in their methodology. Werner was a famous teacher at a mining school in Saxony. The primary emphasis of his work was on mineralogy, and he sought to make an encyclopedic description and classification of the mineral kingdom. Hutton emphasized historical geology and the study of land forms. Simply put (too simply for complete accuracy), Werner was more like Linnaeus and Hutton was more like Buffon. The differences in their approaches to natural history were more differences of subject and method than differences of belief.

These differences of method were most clearly spelled out in the German tradition. Leibniz and Christian Wolff had made a careful distinction between the visible world of nature and the ideal, abstract world of the mind. History, in this philosophical tradition, belonged firmly in the visible world of nature, because history could only record a continuous series of actual beings or events. The history of nature involved no arbitrary choice of characteristics and no logical division of forms such as those employed in the taxonomy of Linnaeus. Buffon knew of this distinction in Leibnizian philosophy largely through reading Madame du Châtelet's *Institutions de physique.* Buffon's familiarity with it can be seen in his statement of method in the *Histoire naturelle,* where he condemned Linnaeus's taxonomy for being artificial. Buffon believed that his taxonomy was natural because it was historical.

Kant, the eighteenth-century philosopher who most skillfully unraveled these metaphysical and methodological knots, read Buffon's *Histoire naturelle* and used the same methodological distinctions as early as the 1750s. Kant distinguished between the history of nature (*Naturgeschichte*) and the description of nature (*Naturbeschreibung*). A taxonomy in *Naturgeschichte* need not be artificial, because the reproductive histories of families of individuals define the species without making it necessary for the natural historian to arbitrarily select characteristics. But taxonomy in the realm of *Naturbeschreibung* would of necessity be a logical division imposed by the mind upon nature. In geology, Hutton and Werner ended up on opposite sides of this distinction, Hutton favoring a historical treatment of the earth, and Werner favoring a descriptive treatment. But the fact that the system of characteristics employed in *Naturbeschreibung* was artificial did not make it invalid. Kant did not denigrate *Naturbeschreibung.* He recognized, as Buffon had not, that a taxonomy such as that given by Linnaeus was valuable, in fact essential, even though it was to a large extent arbitrary. The same thing can be said of Werner's mineralogy. A taxonomy of rocks and minerals, combined with careful descriptions of their usual occurrences and formations (Werner called this "geognosy"), may not tell us much about their origins, but it is indispensible for geology.

Werner's attachment to the tradition of *Naturbeschreibung* can be seen from the title of his most famous treatise, *Kurze Klassifikation und Beschreibung der verschiedenen Gebirgsarten* [*A short classification and description of the different mineral assemblages*] (1786). It was possible, of course, to have a plain description of nature that did not attempt a taxonomy, and "plain description" was definitely Werner's primary goal. He chose to describe minerals by their external characteristics, such as color, taste, texture, smell, and hardness, rather than by internal characteristics, such as chemical composition and crystalline structure. His motive was largely pragmatic since for miners it is more important to be able to identify minerals than to be able to classify them. Werner said that "to classify minerals in a system and to identify minerals from their exterior . . . are two different things." He would "rather have a mineral ill classified and well described, than well classified and ill described."[14] From a philosophical point of view, a natural system of classification based on history would be more valuable than an artificial system based on external characteristics, but from the point of view of the practicing mineralogist an artificial system that did not require a prior knowledge of the history of the earth was more valuable.

From our modern perspective, *Naturgeschichte* would seem to have had one important advantage over *Naturbeschreibung:* Its temporal nature would appear to make it more receptive to ideas of evolution. But in fact this was not the case. The historical view of nature was most strongly advocated at the middle of the century by Buffon and at the end of the century by Georges Cuvier (1769–1832), both of them believers in the fixity of the species, and both of them natural historians who did not believe in evolution. The philosophers who advocated or at least seriously considered the idea of the transformation of species were Maupertuis, Diderot, and Lamarck, none of whom had any real interest in history. Their emphasis was on generation and the forces in living matter. Thus one finds history and transformism both appearing in the study of the living world during the eighteenth century, but not together.

The ideas that would be important for the theory of evolution appeared during the Enlightenment, but not the theory itself. What was true of evolutionary theory was true of most fields of biology. They were disciplines almost formed by the end of the century – but not quite. The life sciences had changed greatly. The mechanical philosophy, which had been so successful in the physical sciences during the previous century, had failed in the life sciences, but it had succeeded in destroying Aristotle's methodology. A return to the concepts of substantial forms and final causes was impossible. With the old foundation gone and with the new mechanical philosophy also proving inadequate, a search for new methods of investigation and for new theories was inevitable. What the philosophers of the life sciences found – or rather founded – was the science of biology.

CHAPTER VI

The Moral Sciences

In 1774, Turgot – the same Turgot who wrote the article entitled "Expansibilité" for the *Encyclopédie* – became controller general of France under the new king, Louis XVI. For the first time the most important ministerial position in the kingdom was held by a friend of the philosophes'. France had suffered a series of financial crises that had grown in severity throughout the century. The cause was not a general decline in prosperity but a tax system that made it impossible for the king to tax the real sources of wealth in the kingdom. The financial crisis had at its root a social crisis. The clergy, the nobility, and the *parlements* (traditional judicial bodies that claimed the right to approve taxes) jealously guarded their prerogatives and sought to extend their powers with little thought for the state as a whole. By 1774, France was approaching disaster. The failure of fiscal and social reform at that juncture (and Turgot did fail; he stayed in office for only twenty months) meant that special interests would prevail and that future ministers would be chosen not for their reforming skills but for their ability to borrow money. Fifteen years after Turgot began his ministry, France collapsed in a decade of revolution that eventually brought the needed reforms, but only at the expense of war within and without, and protracted political chaos.

When he took office, Turgot instructed the young king on the problems of his realm:

> The cause of the evil, Sire, goes back to the fact that your nation . . . is a society composed of different orders badly united, and of a people in which there are but very few social ties between the members. In consequence, each individual is occupied only with his own particular exclusive interest and almost no one bothers to fulfill his duties or know his relations to others. It follows that there exists a perpetual war of claims and counter-claims which reason and mutual understanding have never regulated.

Turgot's answer was to rationalize social and political institutions, not on the basis of their historical rights and interests but on the basis of natural rights. In the past, decisions had been made "by examining the example of what our ancestors did in times of ignorance and barbarism." Such a course had merely perpetuated previous errors. According to Turgot, the solution was to develop a new objective science of society, founded on the constants of human nature and the mutual needs of all men and women. He argued that "these rights and interests are not very numerous. Consequently, the science which comprises them, based upon the principles of justice that each of us bears in his heart and on the intimate conviction of our own sensations, has a very great degree of certainty and yet is not at all extensive. It does not demand the efforts of long study and does not surpass the capabilities of any man of good will." What Turgot wished to established was a social science.

In order to effect this, the person whose good will it was most important to obtain was the king, because Turgot, along with most of his philosopher friends, believed that the needed reforms could only be imposed from above. "Men of good will" did not include the *canaille* at the bottom of society, who would never understand the new social science, no matter how few and simple its principles. The king was the key to success. If he were enough of a philosopher he would realize that once a social science was established he would no longer have to make decisions on every minute problem: He would be able to "govern like God by general laws."[1] When the laws of France agreed with the laws of nature, the operations of government would be simple and harmonious.

Turgot's version of the science of man was based on reason and experience rather than tradition – partly because that was what the hour demanded, and partly because Turgot and his friend the Marquis de Condorcet believed that the new social science should be empirical and quantitative. It was Turgot, Condorcet, and their circle who first used the term *social science.*

The Science of Man

A social science requires a science of man that uses methods comparable to those of the physical sciences. The search for a science of man began in the seventeenth century and was closely associated with the new experimental philosophy. Voltaire placed its origin in Locke's *Essay concerning human understanding.* Voltaire admitted that other philosophers had attempted to determine the nature of the soul, but only Locke, he said, "had modestly written its history" – that is, its natural history – from the plain experience of reflection on mental acts. David Hume had wished to create from Locke's analysis a science of human nature. He argued that the science of man was basic to all of the other sciences, because all sciences, even mathematics and physics, came back in the end to human sensations and to human reasoning. Any science required that the phenomena it described be uniform and regular. Hume did not doubt that human nature displayed the same uniformity as the physical world. Its principles and operations remained the same "in all nations and ages," the same motives producing the same actions. Hume concluded that "where experiments [from cautious observations of human life] are judiciously collected and compared, we may hope to establish on them a science, which will not be inferior in certainty, and will be much superior in utility to any other of human comprehension."[2]

The philosophers of the Enlightenment agreed that such a science was possible, but they did not all find its principles in the same place. The Scottish philosophers found their principles in a special human sentiment or intuitive sociability; the utilitarians found them in a collection of the self-interest of individuals that accumulated into a general interest of society; the rationalists found them in a mathematical analysis of human rights. For some the science of man was founded in psychology, the science of mental actions, and for others it was a science of social organization, but in every case it was a science modeled on the methods of the physical sciences, which had already proved their effectiveness.

A science of human nature would make it possible to rationalize social institutions, and therefore the creation of social science was a crucial part of the philosophes' program to reform society. The program was already suggested in a few works that had appeared during the first half of the eighteenth century – Pierre Bayle's *Historical and critical dictionary* (1697), Montesquieu's *Persian letters* (1721), Hume's *Treatise on human nature* (1739), and Voltaire's *Philosophical letters* (1735) – but around 1750 there appeared in France a spate of books that suggested an increase in the tempo of philosophical criticism and a greater degree of cohesion among the authors. Montesquieu's *Esprit des lois* [*Spirit of the laws*] appeared at the end of 1748 and was available in France the next year. In 1749 there also appeared Diderot's *Lettre sur les aveugles* [*Letter on the blind*]; the first three books of Buffon's *Histoire naturelle* [*Natural history*,] including the famous preface on the writing of natural history; and Abbé Condillac's *Traité des systèmes* [*Treatise on sys-*

tems]. In 1750 there appeared Turgot's *Discours sur les progrès succésifs de l'esprit humain* [*Discourse on the successive progress of the human spirit*] and Jean-Jacques Rousseau's *Discours sur les sciences et les arts* [*Discourse on the sciences and the arts*]. The first volume of the *Encyclopédie,* containing d'Alembert's *Discours préliminaire* ["Preliminary discourse"], appeared in June 1751 and was soon followed by Voltaire's *Siècle de Louis XIV* [*Age of Louis XIV*]. All of these works have since been recognized as major documents in the history of the Enlightenment, and coming together as they did at one time in a single country they signaled the beginning of a distinct intellectual movement.

It would be a mistake, however, to assume that the philosophes all agreed with one another. Diderot, d'Alembert, Condillac, and Rousseau, who had been close friends in Paris, were all going their separate ways by the end of the 1750s. D'Alembert and Condillac remained loyal to a science built on mathematics and quantitative experiment; Diderot moved towards dynamic materialism; and Rousseau created the first stirrings of Romanticism. Nevertheless there were certain common characteristics in their thought. Most particularly they believed that human actions should be regulated by nature and not by precepts taken from the Bible, and they believed that natural science gave insights into the workings of human nature.

Montesquieu's *Spirit of the Laws*

Montesquieu opened his *Spirit of the laws* with a definition of law: "Laws in their most general signification, are the necessary relations derived from the nature of things. In this sense all beings have their laws, the Deity has his laws, the material World its laws, the intelligences superior to man have their laws, the beasts their laws, man his laws." As the necessary relations among things, laws were not arbitrary. Even the laws of men had their roots in the constants of human nature. As a young man Montesquieu had contributed scientific papers to the Academy of Bordeaux on the causes of the echo, the functions of the kidneys, the principles of weight, the nature of tides, and on fossil oysters. These scientific interests reappeared in the famous fourteenth book of the *Spirit of the laws* in which Montesquieu argued that human temperament varied according to climate and that laws should therefore be expected to vary too. Temperament is a function of the tension or lassitude of the "fibers" that compose the body. In a warm climate the fibers are loose, and humans are relaxed but sensitive. In a cold climate the fibers contract. The individual exhibits greater energy but loses sensitivity. "The inhabitants of warm countries are, like old men, timorous; the people in cold countries are, like young men, brave." The inhabitants of warm climates are emotional and quick to sense pain, but "You must flay a Muscovite alive to make him feel."

Montesquieu learned his physiology from Hermann Boerhaave's *Institutiones medicae* [*Institutions of medicine*] (1708) and from a book by John Arbuthnot (1667–1735) entitled *An essay concerning the effects of air on human bodies* (1733), which discussed the physiological effects of climate and employed the popular fiber theory of the human body. Montesquieu also performed his own experiments. He froze a portion of a sheep's tongue and observed that the papillae on the tongue had diminished in size. "In proportion as the frost went off, the papillae seemed to the naked eye to rise, and with the microscope the miliary glands began to appear. This observation confirms what I have been saying, that in cold countries the nervous glands are less spread; they sink deeper into their sheaths, or they are sheltered from the action of external objects. Consequently they have not such lively sensations."[3] Such experiments confirmed his belief that natural law extended to human behavior. From his observations of men's actions throughout human history, Montesquieu began his search for the "spirit" of the laws, those aspects that relate the character of a people to its legal system and social structure.

He distinguished among three primary kinds of governments – democracy, monarchy, and despotism – each of which was motivated by a different principle. Democracy was based on the principle of civic virtue, monarchy on the principle of honor, and despotism on the principle of fear. The book was not ideologically neutral. Montesquieu's attacks on slavery, despotism, and much of criminal law revealed his liberal leanings, but his intention had been to write a descriptive and comparative study of law. He was a past master of the comparative method, having already compared the faults and virtues of French society with those of Persia in his *Persian letters* (1721). His purpose, he said, was to draw his principles not from his prejudices but from the nature of things, and to describe rather than to judge. He said that he did not wish to criticize whatever was established in any country, yet in spite of his professed objectivity his commitment to liberal democracy is apparent throughout the book.

The *Spirit of the laws* was a prodigious effort that was better received outside of France than in it. D'Alembert wrote to the Swiss mathematician Gabriel Cramer (1704–52) that he did not like the book: "Why try to resolve a problem when some of the necessary 'givens' are lacking? In general I find that all the works like this one of [Montesquieu] rather resemble those physical dissertations . . . where the author explains the phenomena so easily that he could just as well have explained completely different phenomena by the same principles."[4] Voltaire complained with some justification that the book was "a labyrinth without a clue, lacking all method."[5] Condorcet praised Montesquieu's attempt to reveal a unity or "spirit" behind the confusion of custom and privilege that had emerged as law, but he criticized Montesquieu for not deriving from his study a rational system of law. As mathematicians, Condorcet and d'Alembert did not value the historical basis of law. For them the creation of law was a matter of discovering through the science of human nature how people should live. But in spite of these criticisms, the *Spirit of the laws* was probably the most influential single book of the Enlightenment, especially outside of France, and even the French, including d'Alembert, Condorcet, and Voltaire, learned to respect it for its balanced argument and for the great learning that it displayed.

The *Encyclopédie*

The *Encyclopédie, ou dictionnaire raisonné des sciences, des arts et des métiers, par une société de gens de lettres* [*Encyclopedia, or reasoned dictionary of the sciences, arts, and crafts, published by a society of men of letters*] was the focal point for much of the philosophy and reforming spirit at the middle of the century. Early in the seventeenth century Francis Bacon had urged the creation of a great dictionary that would bring together in an orderly fashion all of the practical knowledge that was known only to craftsmen in their respective trades. Efforts by the Royal Society to produce such a dictionary had come to nothing, but the ideal had not been lost. Freemasonry, which began in England and spread rapidly in France during the Enlightenment, was sympathetic to Bacon's philosophy, and the Masons in Paris proposed a dictionary project in 1737. So did the Jesuits. But the encyclopedia that emerged was not obviously associated with either of these groups.

In 1728, an English Quaker by the name of Ephraim Chambers (ca. 1680–1740) published in London a two-volume *Cyclopaedia, or universal dictionary of the arts and sciences* (1728). The *Cyclopedia* was a great success, and so when in 1745 two strangers walked into the office in Paris of the publisher André-François Le Breton with a complete French translation of the *Cyclopaedia,* Le Breton recognized a golden opportunity. He soon quarreled with the translators, however, and the royal "privilege" that had to be acquired before any publication could legally begin was withdrawn. Once Le Breton had extricated himself from legal entanglements with the translators and had reclaimed his privilege, he expanded the project, bringing in other publishers in Paris to provide the large amount of capital needed to finance a ten-volume work.

Le Breton's first choice as an editor proved unsatisfactory, and the job finally fell to Diderot and d'Alembert. Diderot had already been employed as a translator by one of the publishers and had proved his reliability. D'Alembert was also known to the publishers and had the advantage of membership in the Paris Academy of Sciences. He was already a distinguished mathematician, and the task of writing the scientific articles fell to him. In 1749, when Diderot was jailed for his *Letter on the blind,* d'Alembert showed that his commitment to the task was limited. He wrote Samuel Formey in Berlin: "I never intended to have a hand in it except for what has to do with mathematics and physical astronomy. I am in a position to do only that, and besides, I do not intend to condemn myself for ten years to the tedium of seven or eight folios."[6] Diderot's commitment, however, was total. The *Encyclopédie* grew to seventeen volumes of text and eleven volumes of plates, each volume standing 16 inches high and containing over nine hundred double columns of text. The first volume appeared in 1751, and the last volumes came out in 1772. Diderot had persisted through two suspensions of the privilege and constant harassment. D'Alembert gave up his editorship in 1758, after the second and more serious suspension of the privilege, and Diderot had to carry out the rest of the editing and publication clandestinely.

The *Encyclopédie* reflected Diderot's enthusiasm for Bacon's philosophy in several ways. The outline of the work was to follow Bacon's classification of knowledge (see Figure 6.1), and the articles were to cover the trades as well as the sciences. Thus the *Encyclopédie* contained numerous articles on power machinery, forging, mining, ship construction, and every conceivable kind of manufacturing. Diderot, who had been trained as a cutler, or knife maker, by his father, went into the shops of artisans to learn the skills of their trades. The articles on the crafts were illustrated by elaborate copper engravings, some of which were made especially for the *Encyclopédie;* others came from René Réaumur's *Déscription des arts et métiers* [*Description of the arts and crafts*]. (For the plate showing the cutler at work, see Figure 6.2.) As a result, the *Ency-*

clopédie is our main source of information about eighteenth-century technology.

As a "reasoned dictionary," or systematic presentation of knowledge in all the sciences, the *Encyclopédie* provided easy access to information on any conceivable subject – religion, law, literature, mathematics, philosophy, chemistry, military science, and agriculture. As a universal encyclopedia, its purpose was to show the interconnectedness of all knowledge. Its opponents claimed that it was antireligious propaganda intended to destroy the bonds of society, or, more specifically, "to propagate materialism, to destroy Religion, to inspire a spirit of independence, and to nourish the corruption of morals."[7] Diderot and d'Alembert did, of course, try to infuse a reforming spirit into their publication, but in all fairness one would have to admit that they also made an honest effort to obtain the best authors available on each subject. The *Encyclopédie* succeeded not because of its political message but because of the vast amount of accurate information that it contained. It became the model of all subsequent encyclopedias, and its popularity was demonstrated by the speed and frequency with which it was pirated.

The title page announced that the work was written by a "society of men of letters." Voltaire, Montesquieu, Rousseau, and Turgot all wrote for the *Encyclopédie,* although the contributions of these more illustrious members were limited to a few articles. The editors saw the *Encyclopédie* not only as a service to their fellow men but also as a gift to posterity. Diderot wrote in the *Prospectus* (most of which was repeated in the "Preliminary discourse" to the first volume): "Let us hope that posterity will say, upon opening our Dictionary: such was the state of the sciences and the fine arts then. May the history of the human mind and its productions continue from age to age until the most distant centuries. May the *Encyclopédie* become a sanctuary, where knowledge of man is protected from time and from revolutions. Will we not be more than flattered to have laid its foundations?"[8] The *Encyclopédie* was to be a new beginning, a foundation of knowledge, whose constant improvement by succeeding generations would ensure human progress and whose very existence would be a guarantee against ignorance, bigotry, and superstition.

Except for the addition of Diderot's *Prospectus,* d'Alembert wrote the "Preliminary discourse"; it had a great success and made his reputation as a writer. In this essay d'Alembert gave what he called a "reasoned history" of the formation of the sciences. It was not an

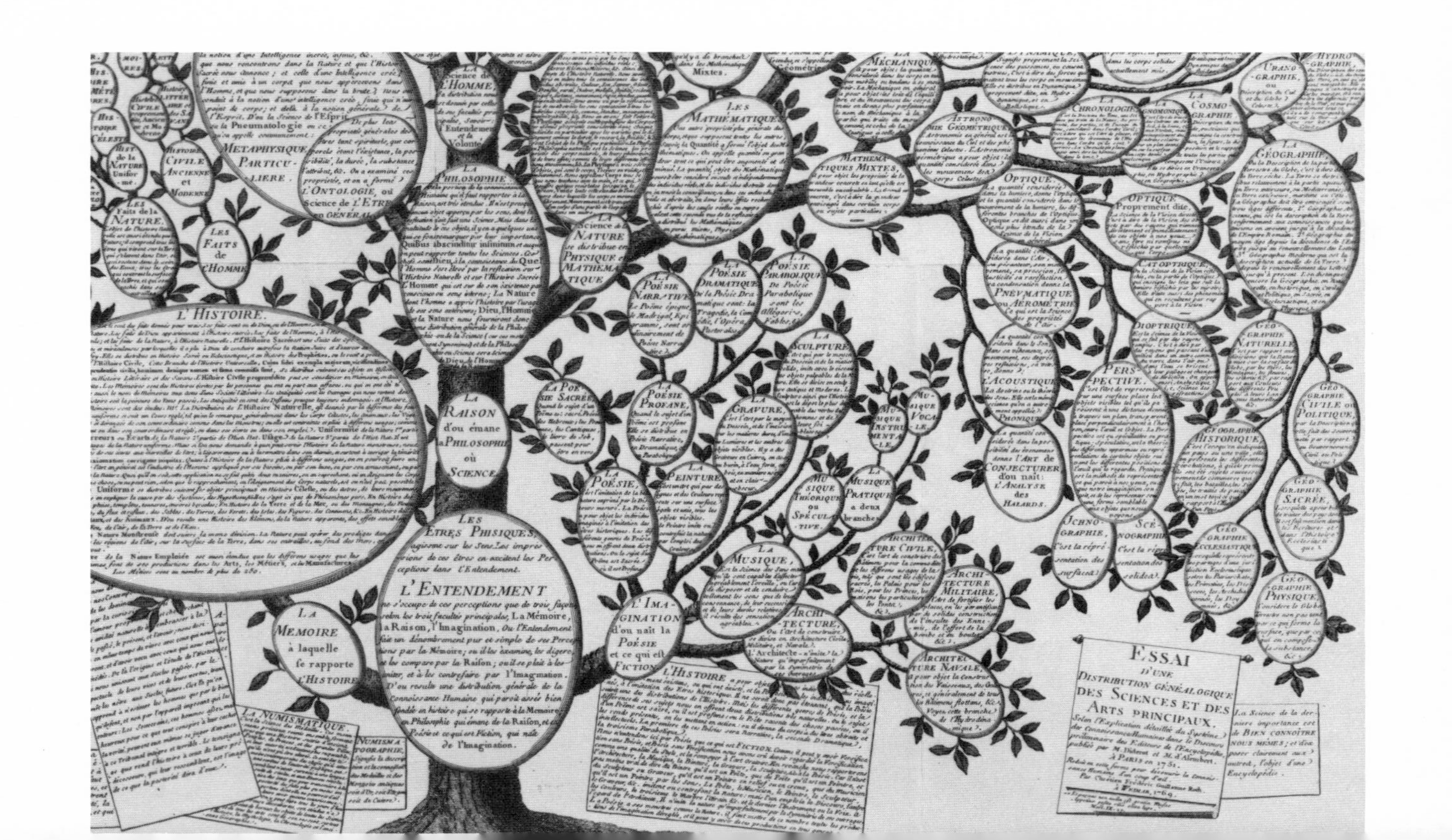

ESSAI
D'UNE
DISTRIBUTION GÉNÉALOGIQUE
DES SCIENCES ET DES
ARTS PRINCIPAUX.
L'ENTENDEMENT
LA MEMOIRE à laquelle se rapporte l'HISTOIRE
LA RAISON d'ou émane la PHILOSOPHIE où SCIENCE.
L'IMAGINATION d'ou nait la POÉSIE et ce qui est FICTION
L'HISTOIRE.
LES ÊTRES PHISIQUES,
LA PHILOSOPHIE
METAPHYSIQUE PARTICULIERE.
LES FAITS de L'HOMME
LES FAITS de la NATURE
HISTOIRE CIVILE ANCIENNE et MODERNE
LA SCIENCE de la NATURE se distribue en PHYSIQUE et MATHEMATIQUE
LES MATHÉMATIQUES
MATHEMATIQUES MIXTES
LA MECHANIQUE
ASTRONOMIE GÉOMÉTRIQUE
OPTIQUE
OPTIQUE Proprement dite.
PNEUMATIQUE ou AÉROMETRIE
ACOUSTIQUE
DIOPTRIQUE
PERSPECTIVE.
ICHNOGRAPHIE
SCÉNOGRAPHIE
LA CHRONOLOGIE
LA GNOMONIQUE
LA COSMOGRAPHIE
URANOGRAPHIE
LA GÉOGRAPHIE
GÉOGRAPHIE NATURELLE
GÉOGRAPHIE CIVILE ou POLITIQUE
GÉOGRAPHIE HISTORIQUE
GÉOGRAPHIE SACRÉE
GÉOGRAPHIE ECCLESIASTIQUE
GÉOGRAPHIE PHYSIQUE
LA POÉSIE
POÉSIE SACRÉE
POÉSIE PROFANE
POÉSIE NARRATIVE
POÉSIE DRAMATIQUE
POÉSIE PARABOLIQUE
LA PEINTURE
LA GRAVURE
LA SCULPTURE
LA MUSIQUE
MUSIQUE THEORIQUE ou SPECULATIVE
MUSIQUE PRATIQUE a deux branches
MUSIQUE INSTRUMENTALE
MUSIQUE VOCALE
ARCHITECTURE
ARCHITECTURE CIVILE
ARCHITECTURE MILITAIRE
ARCHITECTURE NAVALE
LA NUMISMATIQUE

Fig. 6.1. The tree of knowledge, from the *Encyclopédie.* Only a small portion of the tree of knowledge will fit on this page, but it is enough to show that of the three branches of human knowledge – memory, imagination, and reason – reason is the sturdiest. In fact, it is not a branch at all, but the central trunk, which supports the greatest weight. Mathematics branches into "pure mathematics" (not shown) and "mixed mathematics," the latter including mechanics, astronomy, optics, acoustics, probability theory, and even geography. Physics branches off just before mathematics. It still contains botany and zoology, along with chemistry, mineralogy, and meteorology, although these "leaves" of the tree are in the higher branches not shown in this illustration. Natural history, which is founded on the faculty of memory, branches from the opposite side of the trunk. *Sources: Encyclopédie, ou dictionnaire raisonné des sciences, des arts, et des métiers,* ed. D. Diderot and J. d'Alembert, 17 vols. of text (1751–65), 11 vols. of plates (1762–77), 4 suppl. vols. of text, 1 suppl. vol. of plates, and 2 suppl. vols. of index (1776–80) (Paris): vol. I of index, frontispiece. Courtesy of the University of Washington Libraries.

Fig. 6.2. The cutler at work. The chief editor of the *Encyclopédie,* Denis Diderot, was born into a family of cutlers (makers of knives). This engraving from the *Encyclopédie* must have had special significance for him. It also shows why he preferred a literary career to that of a cutler. The reclining position of the cutler at the grindstone was characteristic of the trade. *Sources: Encyclopédie, ou dictionnaire raisonné des sciences, des arts, et des métiers,* ed. D. Diderot and J. d'Alembert, 17 vols. of text (1751–65), 11 vols. of plates (1762–77), 4 suppl. vols. of text, 1 suppl. vol. of plates, and 2 suppl. vols. of index (1776–80) (Paris): vol. III of plates, "Coutelier," pl. I. By permission of the Syndics of Cambridge University Library.

actual history but an account of how the sciences might logically have come into being if they had followed a rational sequence of discovery. Although d'Alembert followed Diderot's lead and emphasized the Baconian structure of the sciences, his reasoned history moved quickly from the first intellectual stirrings of man in his primitive state to the contemplation of abstract bodies that was the foundation of geometry and mechanics. D'Alembert was enough of a Lockean to admit that all knowledge comes from the senses and that the sciences must therefore depend on observation and experiment, but he also believed that once experiment had revealed the basic laws of a science, then the elaboration of theory was the work of mathematicians.

Buffon's *Discourse* criticizing mathematics appeared in 1749 when the *Encyclopédie* was just getting under way. D'Alembert's criticism of it was similar to his criticism of Montesquieu's *Spirit of the laws.* He wrote to his friend Cramer: "You will find us badly treated in the new work by Buffon. It is true that with a few more calculations and a little more geometry he perhaps would not have hazarded so many notions on the formation of the earth."[9] D'Alembert's colleague Diderot reacted quite differently and repeated Buffon's criticisms of mathematics and mechanics in his *Pensées sur l'interprétation de la nature,* a title taken directly from Bacon. As Diderot's interests turned from mathematics to physiology and chemistry, his friendship with d'Alembert weakened. D'Alembert remained the mathematical rationalist that he had always been. He left the *Encyclopédie* in 1758, ostensibly because continuing the work was becoming dangerous, but the philosophical rift opening between the two editors was probably the major cause of their separation.

When d'Alembert resigned, Diderot complained to Voltaire:

> The reign of mathematics is over. Tastes have changed. Natural history and letters are now dominant. D'Alembert at his age will not throw himself into the study of natural history, and it will be difficult for him to write a literary work that will live up to the celebrity of his name. Several articles in the *Encyclopédie* would have sustained him with dignity. . . . This is something that he has not considered, that perhaps no one dares tell him and that he will hear from me, because I am made to tell the truth to my friends.[10]

Undoubtedly this was a "truth" that d'Alembert did not like to hear, but then neither did Voltaire.

In the 1760s d'Alembert's friendship with Voltaire became stronger as his friendship with Diderot weakened. D'Alembert had become Voltaire's chief correspondent and his first "lieutenant" in the battle that Voltaire waged for philosophy from his mountain retreat in the Alps on the Swiss border, where he could easily escape arrest in case his fiery attacks against the church inflamed the French authorities into action against him. Diderot became more closely associated with Baron d'Holbach, whose *Système de la nature* [*System of nature*] (1770) became known as the "atheists' Bible." D'Alembert called the book a "detestable stupidity . . . prejudicial both to kings and to the *philosophes.*"[11] It was objectionable both politically and philosophically. The materialists, argued Voltaire, were forced to say

> . . . that the material world has thought and sentiment essential to itself, because [without a God] it has no way to acquire them. . . . This thought and this sentiment would have to be inherent in matter as are extension, divisibility, and capacity for motion. . . . After being led thus from doubt to doubt and from conclusion to conclusion [we are able] to regard this proposition, *"There is a God,"* as the most probable that man can hold . . . after having seen that the contrary proposition is one of the most absurd.[12]

D'Alembert was a skeptic in religion and would not go along with Voltaire's deism, but he agreed with Voltaire that thought and sentiment could not be essential properties of matter as the materialists seemed to require. The course of Diderot's philosophy was one that d'Alembert and Voltaire could not follow.

In spite of the differences between the editors, the *Encyclopédie* welded together the "society of men of letters" into a recognized party. Anyone who took part in the enterprise or was in sympathy with it was given the name *"encyclopédiste."* The *Encyclopédie* was the central document of the Enlightenment. It proved that the philosophes were more than carping critics; it was their positive program for enlightenment and human progress. In the wake of the intellectual ferment of the 1750s, the *encyclopédistes* turned from philosophical reflection to political action, at least to the extent that that was possible in an absolute monarchy.

Scientific Academies

After resigning as editor of the *Encyclopédie,* d'Alembert began a campaign to capture the academies for philosophy. Scientific academies and societies were especially important in the eighteenth century because the universities were not receptive to the teaching of science and even less so to scientific research. The academies gave position and status to scientists who would otherwise have had no place in a highly structured corporate society. This was particularly true of those who were self-taught or whose subjects were far removed from medicine or from the traditional university curriculum. The two great national academies, the Royal Society of London (founded 1662) and the Paris Academy of Sciences (founded 1666), were the models on which the new academies of the eighteenth century were founded. The Paris Academy was a small, exclusive group of professionals salaried by the state. The Royal Society of London was a large group of amateurs who paid dues to support their society's activities.

The Berlin Academy of Sciences was founded in 1700 but achieved little until it was reorganized in 1743 on the Parisian model by Frederick the Great. Frederick, who admired French literature and philosophy, required that the scientific papers at his academy be written in French, and he brought in Pierre Maupertuis as president. He attracted several stars, notably the mathematician Leonhard Euler, by providing reliable salaries that were generally higher than those in Paris. Still higher were salaries at the Russian Academy of Sciences in St. Petersburg, founded by Peter the Great in 1724. Many members of the Russian Academy were recruited from Germany and Switzerland. Euler was at St. Petersburg both before and after his years at Berlin. Other prominent members were Daniel Bernoulli, George Bernhard Bilfinger (1693–1750), Georg Wolfgang Krafft (1701–59), and Georg Wilhelm Richmann, whose contributions to the study of electricity and heat we have already noted. Other royal academies were founded in Göttingen (1751), Bologna (1714), Turin (1757), and Munich (1758). These smaller academies were less well endowed than the great national academies, but they published scientific memoirs in most cases and provided at least some support for their members.

The Paris Academy of Sciences and those academies modeled after it were required to serve the state in a variety of capacities. These obligations could become burdensome, especially at Paris, where the academy was expected to act as a patent office and office of censorship as well as a government research laboratory. The members of the Royal Society of London, because they did not receive salaries, were free from obligations to the state. Societies modeled after the Royal Society were founded at Edinburgh (1783), Manchester (1781), Haarlem (1756), and in many of the French provinces. The Swedish Academy of Sciences at Stockholm began as an amateur academy in 1739, but it received several large bequests and the income from an extremely profitable almanac business, which allowed it to grant pensions to its members. (The publication of almanacs was often the monopoly of an academy. The Berlin Academy also received substantial income from the sale of almanacs.)

Besides publishing the research of their members and of foreign correspondents, the academies also held prize competitions on a wide variety of subjects. Many prizes were given for the solution of problems in theoretical and applied mechanics. Prize competitions on the action of sails, anchors, capstans, and the solid of least resistance in a fluid were obviously proposed to benefit the navy and the merchant fleet. Other prizes on philosophy and literature were offered by the academies that were formed to study those subjects. Of all the prize papers, Rousseau's *Discours sur les sciences et les arts* [*Discourse on the sciences and the arts*], written for the prize offered by the Dijon Academy in 1750, has had the most lasting fame. Another extremely important prize competition was the Paris prize competition of 1720 on the collision of hard bodies, which was repeated over several years and was the focus for a debate over the nature of force, matter, and mechanical action. The prize papers were always submitted anonymously, but a prominent scientist could usually count on the judges recognizing his paper. The prizes were supposed to be on timely topics and were often used to smoke out a competitor, as when Euler persuaded the Berlin Academy to choose the problem of describing the motion of the moon in order to attract a paper on the subject from Clairaut.

In France the royal academies, especially the Academy of Sciences and the literary French Academy, were an integral part of the ancien régime. To capture the academies would be to capture the official intellectual center of the kingdom. It was one way to make the philosophes respectable. D'Alembert had achieved literary fame with his "Preliminary discourse" to the *Encyclopédie.* At first he protested the efforts of his friends to get him into the literary French Academy, but in 1754 he was elected through the machinations of his powerful patroness the Marquise du Deffand (1697–1780). He was, of course, already a member of the Academy of Sciences, having been admitted as an adjunct member in 1741, but his promotion in that academy had come slowly for him, and he was not made *pensionnaire* (with a substantial pension) until 1765. In 1772 he became secretary of the literary French Academy, and from that powerful position he was able to bring in enough "philosophical" members to obtain a majority of the forty seats. Victory in the Academy of Sciences did not come until 1777, when he obtained the position of secretary for his protégé Condorcet, over the strong objections of Buffon. D'Alembert declared that even squaring the circle would not have made him happier.

In spite of the progressive character that d'Alembert and Condorcet tried to bring to the academies, they remained elite institutions that jealously guarded their positions. In the 1760s and 1770s, new societies sprang up throughout France. Most were professional societies such as the new academies of medicine, surgery, and pharmacy. There were also numerous agricultural societies, museums, and societies of artisans. The Academy of Sciences managed to scuttle any of these societies that it regarded as competitors. Most important were the *Société des Arts* and the *Société Libre d'Émulation,* both societies of artisans devoted to the promotion of technology. Throughout the eighteenth century there had been vociferous protests against the Academy of Sciences by artisans who believed that the academicians had suppressed their creativeness and had robbed them of their inventions. The problem was an inevitable result of the functioning of the academy. According to the regulations of 1699, the academy was to examine, "if the King so rules, all machines for which a privilege has been requested from His Majesty. It will certify whether or not they are new and useful, and the inventor whose work has been approved will be held responsible for leaving the Academy a model [of his invention]."[13] By becoming the arbiter in scientific and technological matters, the academy invited the hostility of all those whom it had disappointed. The craftsmen wanted to be judged by their peers, not by an elite group of scholars whose sole purpose seemed to be to act as a barrier between them and the king.

In 1793, during the French Revolution, all of the royal academies in France were suppressed. Some members of the revolutionary Convention that produced the decree sought an exception for the Academy of Sciences, because of its utility. But the elitist and monarchist image of the academy was too strong. The Academy of Sciences disappeared with the rest, and its strongest defenders in those last hours, Lavoisier and Condorcet, lost their lives in Robespierre's Terror.

Jean-Jacques Rousseau, Critic of Society

The science of man took several very different directions during the Enlightenment, the most controversial being that of Jean-Jacques Rousseau (1712–78). Rousseau wrote his prize essay of 1750 in answer to the question "Has the Renaissance in the sciences and the arts contributed to the purification of morals?" Rousseau's answer was no. Since that answer ran counter to the Enlightenment belief in the inevitability of progress through reason, the essay met with vigorous protests. Rousseau had discussed the subject with Diderot in the dungeon at Vincennes, where Diderot had been imprisoned for writing his *Letter on the blind.* Diderot had urged him to take the negative side of the argument, because that was the unexpected answer. But Rousseau became completely committed to a point of view that Diderot must have regarded only as a favorable debating position.

Rousseau found the proper basis for social relations in virtue. Truth became as much a matter of the heart as it had been a matter of the mind. Rousseau's works were full of passion. In his first discourse (1750), he wrote: "Our souls have been corrupted to the degree that our arts and sciences have advanced toward perfection." In the *Discours sur l'origine de l'inégalité* [*Discourse on the origin of inequality*] (1755), he wrote:

> The first man who, having enclosed a piece of land, thought of saying, *This is mine,* and found people simple enough to believe him, was the true founder of civil society. How many crimes, wars, murders, miseries and horrors, might mankind have been spared, if someone had pulled up the stakes or filled in the ditch, and shouted to his fellow-men: "Beware of listening to this imposter; you are ruined if you forget that the fruits of the earth are everyone's and that the soil itself is no one's."

Emile (1762), his book on education, begins: "All is good as it leaves the hands of the author of things. All degenerates in the hands of men," and this same sentiment opens the *Contrat social* [*The social contract*] (1762): "Man is born free, and everywhere he is in chains."[14] In every case Rousseau's message was that man is born in a state of virtue and that society has corrupted him.

This is not the message the philosophes wanted to hear. They agreed that bad society made bad men, but they were committed to the idea that the socializing forces of reason would inevitably bring progress, not corruption. After reading the first two discourses, Voltaire wrote to Rousseau:

> I have received, Sir, your new book against the human race. I thank you for it. You will give men pleasure by telling them well-deserved truths, but you will not correct them. . . . Never has anyone used such wit to reduce us to animal stupidity. One feels like walking on all fours while reading your work. However, as it is now more than sixty years since I have lost this habit, I feel unfortunately that it is impossible for me to resume it, and I leave this natural gait to those more worthy of it than you or I.[15]

Voltaire suspected that the earliest man was more likely to have been brutish than virtuous.

Rousseau quite accurately replied that he had never said that civilized man ever should or could go back to the jungle. He was merely trying to identify the source of human strife and misery.

Rousseau had not intended his attack on science and society to be an attack on reason. In the *Discourse on the origin of inequality* he stated that "the establishment and the abuse of our political societies . . . can be deduced from the nature of man by the sole light of reason and independent of the sacred dogmas which give sovereign authority to the divine right of kings," and in the *Social contract* he attempted an almost geometric analysis of the body politic.[16] It was through reason that he hoped to reveal human nature for what it was and show how society had artificially constrained it.

Rousseau's insistence on using the "sole light of reason independent of the sacred dogmas" was shared by all of the philosophes. In freeing men's minds from religion, they also wished to free their passions. Rousseau was not the first to extol the passions. Hume recognized that a science of man could not always speak of reason, and he admitted the ultimate primacy of the passions: "We speak not strictly and philosophically when we talk of the combat of passion and of reason. Reason is, and ought only to be the slave of the passions, and can never pretend to any other office than to serve and obey them." The philosophes knew that the passions were the driving force behind human actions and that reason could only direct and restrain but never eliminate them. A "reasonable" man would have to recognize this fact. Hume attacked as inhuman such behavior as "celibacy, fasting, penance, mortification, self-denial, humility, silence, solitude, and the whole train of monkish virtues." They "stupefy the understanding and harden the heart, obscure the fancy and sour the temper."[17] Baron d'Holbach claimed that "to prohibit men their passions is to forbid them to be men."[18] Nor did the philosophes believe that all of the sins condemned by Christianity were truly evil, except in excess. Hume spoke favorably of pride, and Diderot was certain that lust could be a positive virtue. In his *Supplément au voyage de Bougainville* [*Supplement to the voyage of Bougainville*] (written in 1772), Diderot admired the innocent and guiltless sexuality of the Tahitians and contrasted it with the behavior of the sexually repressed young chaplain on the voyage, who, after a brief struggle with his conscience, fell into the welcoming arms of his host's voluptuous daughter, crying: "My religion, my holy orders!" The next day, after listening to the chaplain's description of Christian ethics, Orou, the host, replies:

> You are mad if you believe that there is anything in the universe, high or low, that can add or subtract from the laws of nature. Her eternal will is that good shall be chosen rather than evil, and the general welfare rather than the individual's well-being. You may decree the opposite, but you will not be obeyed. By threats, punishment and guilt, you can [only] make

Fig. 6.3. The passions. The passions, as described by Buffon in his *Histoire naturelle* [*Natural history*]. The drawings represent envy, happiness, scorn, terror, and sadness, but not in the numbered order given in the illustration. The reader may enjoy trying to match the different passions with the expressions illustrated. *Sources:* Georges-Louis Leclerc, Comte de Buffon, *Histoire naturelle, générale et particulière, avec la description du cabinet du roy,* 44 vols. (Paris, 1749–1803). This illustration comes from the section of the *Histoire naturelle* entitled "De l'homme" in the chapter entitled "L'age virile." By permission of the Syndics of Cambridge University Library.

more wretches and rascals, more depraved consciences and more corrupted characters. . . . Your society . . . can't be anything but a swarm of hypocrites who secretly trample the laws under foot, or a multitude of wretched beings who serve as instruments for inflicting willing torture upon themselves; of imbeciles in whom prejudice has utterly silenced the voice of nature, or ill-fashioned creatures in whom nature cannot claim her rights.[19]

Such was Diderot's view of any system of ethics that denied to the passions their proper place in human nature.

The Physiocrats

Another group of reformers, called the physiocrats, or *économistes,* believed that the improvement of society could be brought about by making economic activity agree more closely with the laws implanted in nature by Providence. History had no value for them. They saw in the past only bad examples to be corrected in the future by reason. Turgot, as controller general of France, was the member of this group who had the greatest opportunity to put their doctrines into practice. His advice to Louis XVI to bring the laws of the realm into agreement with the laws of nature was merely a restatement of the major principle of *physiocratie* (meaning "rule by nature"). The leader of the physiocrats was François Quesnay (1694–1774), physician-in-ordinary to the king, who resided in the royal palace of Versailles and enjoyed the patronage of the king's mistress Madame de Pompadour. Other members of the school were Victor Riqueti, Marquis de Mirabeau (1715–89), Pierre Paul Mercier de La Riviere (1720–94), and Pierre Samuel Du Pont de Nemours (1739–1817). Du Pont de Nemours coined the term *physiocratie* in his *Physiocratie, ou constitution naturelle du gouvernement le plus avantageux au genre humain* [*Physiocracy, or the natural constitution of government most advantageous for the human race*] (1768). He served the French government in a variety of posts, survived the Terror, and came to the United States in 1799, where he prepared a plan for a system of national education for Thomas Jefferson. He also served as Jefferson's emissary to Napoleon in 1802 in negotiations that led the following year to the purchase of Louisiana. His son Eleuthère Irénée (1771–1834) founded the Du Pont Powder Company in Delaware, where Du Pont died in 1817.

Quesnay first presented his theories in two articles in the *Encyclopédie,* "Fermiers" ["Farmers"] (1756) and "Grains" (1757), and laid down a more consistent statement of principles in his *Tableau économique* [*Economic picture*] (1758–9). The physiocrats enjoyed their greatest influence during the 1760s. Turgot's fall from power as controller general in 1776 signaled the decline of *physiocratie,* and it was replaced in large part by the doctrines of Adam Smith, (1723–90), whose *Wealth of nations* also appeared in 1776.

The theory of the physiocrats divided the population of France into three classes. The productive class, which was composed of farmers, miners, and fishermen, directly reaped the fruits of the earth. The proprietary class owned the land and provided capital. The artisan or "sterile," class was composed of those who manufactured and distributed goods made from the raw materials produced by the productive class. According to the physiocrats, in an ideal state the productive class should represent one-half of the population, and each of the other classes should represent one-quarter of the population. Quesnay argued that although men did not enjoy equal abilities, they did have the same natural rights, and it was therefore incumbent on the government to allow individuals to pursue their own best interests to the extent that they could without infringing on the natural rights of others. In particular, individuals should be allowed to enjoy the results of their own labor, and the physiocrats therefore asserted the sanctity of property and the necessity of a free market.

The physiocrats wished to simplify the system of taxation by instituting a single tax on the land. They argued that the proprietary and artisan classes merely transformed and transferred the raw material that had originally come from the land. The activities of these classes were useful, but only the productive class added to the real wealth of the nation. The prosperity of the country therefore depended entirely on the size of the "net product," which was the difference between the value of the produce from the land and the cost of its production. Because all taxes ultimately had to come one way or another from this net product, the physiocrats recommended that taxes be imposed as direct taxes on land, rather than being distributed over various parts of the chain of production ranging from raw materials to finished products.

Voltaire gleefully satirized this system of taxation in his *L'homme aux quarante écus* [*The man with forty ecus*] (1768), in which the rich merchant went untaxed while the poor peasant groaned under the burden of the *impôt unique,* the single tax on the land. Voltaire realized, as did most of his contemporaries, that the doctrines of the physiocrats, as they were expressed in their extreme form by Mercier de La Riviere, were hopelessly idealistic, but he did not find fault with their demand for a rational reform of the tax system. The dangerous financial state of France meant that a science of economics was the most urgently needed of all the social sciences.

Probability Theory and the Science of Man

Freeing human nature from the artificial constraints of religion and society was one way to promote the science of man. Another way was to quantify the process of judgment. Descartes had believed that knowledge about the physical world could be deduced rigorously from first principles. But the English philosophers, especially Locke, had countered Descartes's arguments with the observation that the actual judgments that men make in living are not built on certain knowledge but on probable knowledge drawn from experience. Since Locke believed that all knowledge came originally from the senses, he could not accept Descartes's claim that absolutely clear and distinct ideas are innate in men's minds, placed there by God. Locke acknowledged that we could obtain such "scientific knowledge" from the relations among ideas, as in mathematics, where we can perceive agreements and disagreements among ideas of number and figure, but he was forced to conclude that "how far soever human industry may advance useful and experimental philosophy in physical things, scientifical [knowledge] will still be out of our reach." Without the "broad daylight" of certainty and demonstration, man must make do with probable knowledge. "In the greatest part of our concernments," wrote Locke, God "has afforded us only the twilight, as I may so say, of probability, suitable, I presume, to that state of mediocrity and probationership he has been pleased to place us in here."[20] If man were like God, he would act from certain knowledge, but as a fallen being he is left with only probable knowledge. It is from probable knowledge that he must make all the numerous decisions of mundane life. Hume's attempt to construct a science of human nature accepted this same inescapable conclusion that man acts from probable knowledge derived from experience. Thus the study of probability would have to be an essential part of the science of man.

The mathematical study of probability had begun in 1654 in a correspondence between Pascal and Pierre de Fermat (1601–65) over the question of how to divide the stakes of an unfinished game fairly. Huygens expanded the theory of games in his *Tractatus de ratiociniis in aleae ludo* [*Treatise on reckoning in games of chance*] (1657) by asking how one should calculate the expectation in any game of chance. He defined expectation as follows: "One's Hazard or Expectation to gain any Thing is worth so much, as, if he had it, he could purchase the like Hazard or Expectation again in a just and equal game."[21] The problem with this definition is that it defines "expectation" in terms of a "just and equal game," but Huygens had already defined a "just and equal game" as one in which the players have equal expectations. On the face of it, the definition appears to be circular.

Throughout the Enlightenment, mathematicians had difficulty proving that the laws of probability applied to nature. The laws were considered valid if they agreed with the decisions of a "reasonable man," but it was easy to differ over what was reasonable. Even at the end of the century, Pierre-Simon Laplace, who systematized and elaborated the entire theory, wrote in his *Essai philosophique sur les probabilités* [*Philosophical essay on probability*] (1814): "It is seen in this essay that the theory of probabilities is at bottom only common sense reduced to calculus; it makes us appreciate with exactitude that which exact minds feel by a sort of instinct without being able ofttimes to give a reason for it."[22] The ambiguity in the notion of a "reasonable man" made it very difficult to answer the questions "How do men make judgements?" and "How should men make judgements?" The test of the theory seemed to be alarmingly subjective.

Probability as human judgment, not as a mathematical science, had long been important in law and in commercial transactions. Contracts, life insurance, marine insurance, annuities, and money lending all involved balancing risk against expectation. They had been important commercial activites since the Renaissance and were not merely matters of philosophical speculation. These so-called aleatory contracts were part of contract law and were included by Hugo Grotius (1583–1645) in his famous studies of the law during the seventeenth century. They were also important as a way of getting around the Christian prohibition against usury, or money lending at interest. If it could be demonstrated that both borrower and lender had equal expectation from the transaction, then loaning money at interest would not be usurious.

Setting premiums for insurance and annuities, however, had been more a matter of experienced guesswork than a matter of calculation. The cost of insuring a ship and its cargo, for instance, would not be calculated from statistics on past claims but from the insurer's judgment of the seaworthiness of the ship, the danger of the voyage, and the reliability of the crew. The use of statistics to set premiums came only in the 1660s, at the same time as probability theory. The first mortality tables from which life insurance premiums could be calculated were constructed just at the end of the seventeenth century.

Probability theory entered into law not only in the making of contracts but also in the determination of guilt and innocence.

Whether the outcome of a trial was correct or not depended on testimony, and testimony gave only probable knowledge. The problem of reaching true conclusions from probable knowledge had a long tradition in law and became the model for classical probability theory in the eighteenth century. Jakob Bernoulli, for instance, distinguished in his *Ars conjectandi* [*The art of conjecture*] (1713) between the "intrinsic" probability of an event itself and the "extrinsic" probability of the testimony describing it. D'Alembert, Buffon, and Willem 'sGravesande all discussed the varying degrees of certainty in human judgments. According to d'Alembert, some truths were known by "evidence" – that is self-evidence – and were absolutely certain. Others were known by "certitude" – that is, by a uniform and repeatable series of observations that gave them a degree of probability so high that they could be held almost as solidly as truths of evidence. The third group of truths were only "probable." D'Alembert placed the truths of history in this category, because they were based only on testimony. There was no way that they could be checked by repeating the events. The distinctions among different kinds of evidence that were important for law were also important in establishing a correct method in the natural sciences. The development of probability theory during the Enlightenment therefore served a double purpose, making contributions to both civil law and natural law.

Legal Reform

Another way in which probability theory entered into the determination of guilt or innocence was in the judicial proceeding itself. This was a new problem during the Enlightenment, and it became a part of the philosophes' campaign for the reform of the criminal justice system. Although the merits of different judicial systems had often been debated in the past, the idea of using mathematics to maximize the probability of a correct verdict was new in the 1760s. In his *Philosophical letters,* Voltaire had praised the British court system because it guaranteed habeas corpus, trial by jury, and forbade torture. This system that the French so greatly admired had, however, become increasingly vicious during the Enlightenment, especially in crimes against property. By the 1760s there were 160 crimes that brought the death penalty in England, more than double the number a century earlier, including pocket picking, forgery, destruction of bridges and canals, and theft. By 1820 the number had risen to 220. The same severity existed in France but with the added threat of torture, and without the protection of jury trial and habeas corpus. Because the sentences were so severe, judges often granted clemency, but usually not to the poor or to religious dissenters.

The campaign of the philosophes to reform the criminal code in France began with the Calas affair. Jean Calas was a Huguenot (Protestant) cloth merchant living in Toulouse. On October 13, 1761, his son Marc-Antoine was discovered hanged in the back of his father's shop. The family at first claimed that Marc-Antoine had been murdered but later said that it was a suicide. The reason they lied at first was apparently because suicide, according to French law, was a felony that required that the corpse be given a mock trial, be dragged naked through the streets, and finally be hanged, to the great humiliation of the victim's family. Testimony was collected from the neighbors, who repeated a rumor that Jean Calas had murdered his son in order to prevent him from converting to Roman Catholicism. The *parlement* of Toulouse sentenced Calas to die at the stake. Because he refused to confess, he was subjected to a variety of tortures and was finally broken on the wheel. Still denying guilt, he was strangled by the executioner.

When Voltaire heard the story in 1762 he was enraged. Everything about the case represented what he hated most. Calas was the victim of religious prejudice; the evidence against him was all circumstantial; the facts of the case (as opposed to the rumors repeated by the witnesses) all pointed to the son's suicide; brutal torture was used in an effort to extract a confession of guilt from the father. All of these procedures were barbaric, and to make matters worse, they had all been carried out in strict accordance with the criminal ordinances of France. Voltaire threw himself into the case and after three years of argument and agitation obtained a reversal of the verdict, too late to help Calas, but not too late to help his family.

Voltaire's campaign on behalf of the Calas family coincided with the appearance of the most important book on the theory of criminal law to be published during the Enlightenment. This was the *Tratto dei delitti e delle pene* [*Essay on crime and punishments*] (1764) by the Marchese di Beccaria (1735–94), which was translated into French in 1766 and went through seven printings. Beccaria said little that was new, but his criticism of existing codes of law was impassioned and cogently argued. He recognized that the science of man was less certain than astronomy or physics, but he nevertheless believed it to be subject to calculation: "It is impossible to prevent all disorders in the universal combat of human passions. They increase in a ratio compounded of population and the conflict of private interests, which it is not possible to turn with geometric precision in the direction of public utility. For mathematical exactitude we must substitute, in the arithmetic of politics, the calculus of probabilities."[23]

Social Mathematics

Beccaria did not attempt to compute these probabilities himself. That task was undertaken by Condorcet. In his acceptance speech in 1782 when he became secretary of the Paris Academy of Sciences, he chose to speak about "those sciences, almost created in our own day, the object of which is man himself, the direct goal of which is the happiness of man." The moral sciences, he believed, followed the same method as the physical sciences and should therefore acquire "the equally exact and precise language, [and] attain the same degree of certainty."[24] His *Essai sur l'application de l'analyse à la probabilité des décisions rendue à la pluralité des voix* [*Essay on the application of analysis to the probability of decisions obtained by a plurality of votes*] (1785) attempted to answer the question "Under what conditions will the probability that the majority decision of an assembly or tribunal is true be high enough to justify the obligation of the rest of society to accept that decision?" In other words, "How can one mathematically maximize the probability of a correct verdict in such a way that it is mathematically advantageous for the citizens?" Condorcet was famous for the obscurity of his mathematical works, but the obscurity of this particular problem of moral or political arithmetic was not his fault. It had been obscure throughout the century. Two famous examples will illustrate the problem.

In the 1730–1 volume of scientific memoirs of the Russian Academy of Sciences, Daniel Bernoulli presented the following paradox in the theory of games. Jean and Pierre agree to play a game in which Jean pays Pierre a fixed sum. Jean then flips a coin. If he throws heads, Pierre pays him one dollar (or one franc, or whatever sum is agreed upon). If Jean does not throw heads until the second toss, Pierre pays him two dollars; if not until the third toss Pierre pays four dollars, the amount owed by Pierre doubling at each toss. The game ends when Jean throws heads. The problem is to determine how much money Jean should pay down at the beginning of the game in order for it to be fair.

The mathematical theory of games said that Jean should pay down an infinite sum because the decreasing probability of throwing a long series of tails is offset by the increasing return and there is no set limit to the number of tosses. But, as Bernoulli pointed out, any gambler would happily play the game for a relatively small sum paid down because experience shows that a very long series of tails never occurs. If the validity of probability theory were determined by its agreement or disagreement with the judgment of a reasonable man, then probability theory failed in the case of the St. Petersburg problem.

There were many attempts to explain the paradox. Bernoulli gave an economic explanation. No gambler can pay down an infinite sum, nor can a gambler flip a coin an infinite number of times; therefore the game is impossible. The moral expectation of a player is also different from the mathematical expectation. A rich gambler can more easily risk a large stake than a poor gambler. Buffon gave a psychological explanation. "Moral" probabilities, which are the judgments made by men in taking risks, depend on the psychology of the individual balancing hope of gain against fear of loss. Not every individual will see the same value in a risk. Bernoulli argued persuasively that the "moral" probability of both players in an even game is negative, because the disadvantage of a loss is always greater than the advantage of an equivalent gain.

D'Alembert took the most drastic position of all. He agreed that the mathematical theory of probability was rigorous and correct, but he denied that it applied to the physical world of nature. The "reasonable man" makes judgments that disagree with the mathematical theory because he knows that the world does not act the way the theory predicts. He believes that it is physically impossible to throw a hundred heads in a row. If it does happen, he concludes that there is something wrong with the coin, not that the mathematical theory is correct. Here is the crux of the problem. One needs random events to test the theory, but at the same time one needs the theory to demonstrate that the events are random. From experience d'Alembert concluded that a coin toss is affected by previous tosses in the series, in direct contradiction to the usual assumption of equiprobability. He warned against any hasty acceptance of mathematical probability in the science of man. "Would it not be astonishing," he wrote, "if the formulas by which one proposes to calculate uncertainty did not participate in the uncertainty themselves . . . and leave some clouds in the mind as to the rigorous truth of the results they furnish?"[25] D'Alembert favored the application of mathematics in physics wherever its use was valid, but as a mathematician he also recognized that there was an unjustified leap in logic in moving from mathematical probability directly to physical events.

The St. Petersburg problem was a theoretical test of probability theory, but inoculation was a real test of social mathematics. Before the discovery of vaccination in 1788 by Edward Jenner (1749–1823), immunization against smallpox could be obtained only by contracting the disease or by "inoculation," which meant taking a pox from a person suffering from the disease and placing it in an incision or scratch in the arm of the person to be immunized. The inoculation would either give the person the disease immediately or it would protect him or her for life. Because the incidence of smallpox was very high, as was the chance of death or disfigurement from the disease, determining the value of inoculation was not an idle speculation. One of Voltaire's *Philosophical letters* had advocated inoculation, and La Condamine, after his return from Peru, championed it as an enlightened medical practice. Inoculation became a popular liberal cause, much praised by the philosophes.

In 1760 Daniel Bernoulli attempted to calculate the change in life expectancy resulting from inoculation in a paper entitled "Essai d'une nouvelle analyse de la mortalité causée par la petite vérole, et des avantages de l'inoculation pour la prevenir" ["Essay on a new analysis of the mortality from smallpox and on the advantages of inoculation to prevent it"]. D'Alembert immediately responded with a criticism that must not have been popular with his fellow philosophes. Setting aside the facts that the information for computing life expectancies was inadequate and that Bernoulli's assumption of equal mortality rates for all ages was highly dubious, d'Alembert attacked Bernoulli's method. A person contemplating inoculation would not think only of his or her life expectancy. The one chance in two hundred of death in the next two weeks would weigh much more heavily with the person than the possibility of adding four years at the end of his or her life. Different portions of one's life may have different values to the individual, and the value of life differs when it is judged by the individual and when it is judged by the state. A correct science of man should agree with "reasonable" judgments, but Bernoulli's calculations did not seem to fit this criterion.

In 1774 Condorcet told Turgot that he had begun to write an essay on probability but that it was a work "more philosophical than mathematical."[26] His friend Pierre Laplace was also beginning to study probability at the same time. Both Condorcet and Laplace were protégés of d'Alembert, and both adopted some of his views, although they soon concluded that d'Alembert's skeptical position was too extreme. Condorcet was fervently in favor of reform and of the creation of a new social science. Laplace, the consummate mathematician, continued to extend the mathematical theory. He eventually recast it in analytic terms and extended its mathematical methods in his *Théorie analytique des probabilités* [*Analytical theory of probability*] (1812).

Condorcet's hopes for a social science were greatly stimulated by a paper that Laplace wrote in 1774 entitled "Mémoire sur la probabilité des causes par les événements" ["Memoir on the probability of causes from the events"]. Condorcet, as assistant secretary of the Paris Academy of Sciences, rushed the paper into print. He wrote in his introduction to the volume of *Mémoires* in which Laplace's paper appeared: "It is clear that this question comprises all the applications of the doctrine of chances to the functions of life and that it is the only useful part of this science, the only part worthy of serious cultivation by Philosophers."

The new subject that so excited Condorcet was the inverse theory of probabilities. It was in fact not new but had been pursued more diligently in England than in France. Ordinary probability theory applies primarily to games of chance. Given the chance that an event will occur (one-half that a coin will come up heads, one-sixth that a die will come up deuce), the theory calculates the probability of success in any given game. From given initial conditions one calculates the probability of events. In the inverse theory the events are known, and one attempts to calculate from their frequency the rules for their occurrence and their probable causes. Jakob Bernoulli had made a beginning on this problem in his *Ars conjectandi* (1713) with his "law of large numbers," which stated that the probability of an event is ever more closely approximated by a larger number of instances. Abraham de Moivre's *Doctrine of chances* (1718), the next most important book on probability after Bernoulli's, extended the theory of inverse probabilities with the intention of showing that the apparent randomness of events in nature will, if subjected to calculation, reveal an underlying order expressing "exquisite Wisdom and Design."[27] This was an important problem in England, where a strong commitment to natural theology was combined with a distrust of a priori arguments in science. Hume's argument that we perceive only events and therefore can have no direct knowledge of causes exposed the weakness in the supposed proof of God from his design, but it made the calculation of inverse probabilities all the more important, because inverse probabilities could reveal physical laws from the uniformity and frequency of events.

The Rev. Thomas Bayes (1702–61) devised an important theorem, published in 1764, that was essentially the same as Laplace's theorem of 1774. Bayes discovered how to calculate from the frequency of an event the chance that it will fall between any two given degrees of probability. It was the most important step in the theory of inverse probability, and it led some to believe that the mathematicians had at last found an answer to Hume's problem of how to determine causes.

Laplace approached the problem of causes as a strict determinist. He believed that probability was nothing real, nor was it a fundamental law of nature: It was merely an expression of our ignorance. Because we do not know all the real, mechanical causes or things, we have to resort to conjecture, and some conjectures can be shown mathematically to be more plausible than others. Inverse probability opened up opportunities not only for a social science but also for physics in the process of induction from experiment and in estimating error. This promising field of statistics was exploited by Laplace especially in celestial mechanics.

Condorcet sought to use these new discoveries to create a social science – one that would lead to a collective discovery of truth and not merely to the domination of the will of the majority over that of the minority. If one could demonstrate mathematically how to maximize the probability of a correct decision by a tribunal or legislature, then the dictates of reason would enforce consent by the citizens to the decisions of that tribunal. Truth would be served not merely because the decisions were the will of the majority but because they maximized justice. Condorcet was more optimistic than Rousseau, who had theorized that man consented to the general will only because truth in social matters was unattainable.

Condorcet's social science was not sociology in its modern sense, because it began with the rights, wills, and decisions of individuals rather than with the observed behavior of societies. The social thinkers of the Enlightenment were reformers who wished to discover the laws by which society should be governed, rather than the laws that it actually followed.

Because of his emphasis on the individual, Condorcet was drawn inevitably to the conclusion that the most enlightened decisions would be made by the most enlightened legislators, a conclusion in conflict with his ideals of liberal democracy. In fact he reluctantly concluded that "it is clear that it can be dangerous to give a democratic constitution to an unenlightened people." To resolve this conflict, he campaigned for educational reform at the same time that he campaigned for political reform, but he could never satisfactorily resolve the tension between scientific elitism and democratic liberalism in his thought.

The End of the Enlightenment

The ideas of the philosophes regarding the science of man were carried into the French Revolution, but not by the philosophes themselves. Almost all of the famous figures of the Enlightenment were gone by 1789, when the revolution broke out. Diderot had died in 1784, d'Alembert in 1783, Turgot in 1781, Condillac in 1780, Voltaire and Rousseau in 1778, Hume in 1776, and Montesquieu in 1755. Only Condorcet, among the leaders of the Enlightenment, was left to carry their ideals into the revolution.

As the financial crisis of France deepened, Condorcet redoubled his efforts to bring his social science into the political arena. He opposed the calling of the Estates-General, because it was an ancient assembly created to resolve the problems of a different society in a different age; but when its convening became inevitable, he fought to bring to it liberal representatives committed to the principles of the rights of man and to rational solutions of the country's problems. Throughout the course of the revolution, Condorcet remained active, first as a member of the Society of Thirty, which worked for the election of liberal deputies to the Estates General, and then as one of the Society of 1789, which formed at the end of 1789 as a forum for the more moderate patriots such as Condorcet and Lafayette. Condorcet attempted to find reasonable solutions and remain above party strife, but that was difficult to do in the midst of a struggle for power. Writing for the Society of 1789, he declared, "We regard the social art as a true science, founded like all the others on facts, experiment, reasoning and calculation; susceptible, like all the others, of indefinite progress and development." The Society of 1789 had admirable goals; but like the Academy of Sciences, with which it had many connections, it was an elitist group of men who believed that they knew what was best for France. As an academy of intellect, it lacked the force of a political party; and its more politically astute members, such as Comte de Mirabeau and the Marquis de Lafayette, joined the Jacobins.

Condorcet carried his reforming zeal into education, a subject that had been dear to the hearts of the philosophes. He wrote five articles on public education in 1791, and in April 1792 he presented a report to the Legislative Assembly on behalf of its Committee on Public Instruction. Not surprisingly, the report urged an education in the physical and social sciences, which it held to be the most useful fields of study for good citizens. By this time the politics of revolution was rapidly undermining any scientific analysis of human nature. Condorcet had called for a republic after the king's attempt to flee the country in June 1791, but even so he was slow to detect the winds of political change. On the very eve of the insurrection of August 10, 1792, Condorcet was lecturing the members of the Legislative Assembly on the principles of representation – a subject that was rapidly becoming meaningless. Under the Convention that followed, he was the principal author of the *Plan for a constitution* presented by the Committee on the Constitution to the Convention on February 15, 1793. It was a complex document embodying the ideals of Enlightenment philosophy and designed to obtain, through proper representation, the collective reason of the greatest number. "I say its reason," wrote Condorcet, "and not its will, because the power of the majority over the minority must not be arbitrary. . . . This distinction is not futile: a collection of men, like an individual, can and must distinguish what it wants from what it finds just and reasonable." Condorcet's appeal to reason was lost in the political struggle for power between the Jacobins and the Girondins. On June 2, 1793, the Parisian mob that Condorcet had always feared rose at the instigation of the radical Montagnards and forced the arrest of twenty-nine deputies. Now firmly in control of the Convention, the Montagnards gutted Condorcet's constitution to serve their own purposes.

Condorcet could not accept the destruction of his most cherished ideas. He protested in an appeal *Aux citoyens français, sur la nouvelle constitution* [*To the citizens of France on the new constitution*]. The constitution of the Montagnards, he said, served the interests of one political party, whereas his constitution had been written to serve a nation. He was denounced in the Convention as a man "who, because he sat with some savants in the Academy, imagines it his duty to give laws to the French Republic!" An order went out for his arrest. For eight months he worked in hiding, preparing an introduction for what he hoped would be his greatest work on the social sciences. This introduction, entitled *Esquisse d'un tableau historique des progrès de l'esprit humain* [*Sketch of a historical outline of the progress of the human spirit*], became the final document of the Enlightenment, and its appeal to posterity. In the midst of brutality, unreason, and chaos, Condorcet reaffirmed the vision of the Enlightenment. The laws of nature remained necessary and constant. They applied equally to the physical universe and to the intellectual and moral faculties of man. Therefore, the progress of the science of man was as inevitable as the progress of physics.

How welcome to the philosopher is this picture of the human race, freed from all its chains, released from the domination of chance and from that of the enemies of its progress, advancing with a firm and sure step in the path of truth, virtue and happiness! . . . This contemplation is for him an asylum where the memory of his persecutors cannot pursue him, where he forgets man tormented and corrupted by greed, fear, or envy, to live in thought with man restored to the rights and dignity of his nature. There he truly lives in communion with his fellows, in a paradise that his reason has been able to create and his love of humanity enhances with the purest of pleasures.

As the tempo of the Terror increased, Condorcet feared that his hiding place would be revealed. He fled Paris disguised as a woman, was turned away by former friends, and was arrested on March 27, 1794, in a country inn. Two days later he was found dead in his cell, apparently from exhaustion.

The period that we call the Enlightenment ended with Condorcet, but its ideals did not die with him. Ironically, the unreason that destroyed him also destroyed the old regime that had resisted the reforms of the philosophes. When Europe emerged from the revolutionary era, the science of man and the science of nature both bloomed again with remarkable vigor.

Bibliographic Essay

The most detailed general study of eighteenth-century science is A. Wolf, *A history of science, technology and philosophy in the eighteenth century,* 2 vols. (New York, 1939). It is especially good on the history of technology and on scientific instruments. For information on any particular science, Wolf's book is a good place to begin. Because of its age, however, it does not contain recent interpretations of Enlightenment science. *Natural philosophy through the eighteenth century,* ed. Alan Ferguson, commemoration number of the *Philosophical Magazine* (London, 1948), is a collection of essays on the different sciences and their cultural setting. Colm Kiernan, in his *Enlightenment and science in eighteenth-century France,* 2d ed. (Banbury, U.K., 1973) associates philosophical differences in the Enlightenment with different views of the physical sciences and the sciences of life. Other valuable essays on the eighteenth century are in Charles Gillispie, *The edge of objectivity, an essay in the history of scientific ideas* (Princeton, 1960). The historiography of Enlightenment science is complex and controversial at present, and for that reason I have not included it in this book. Good discussions of the points at issue can be found in *The ferment of knowledge: studies in the historiography of eighteenth-century science,* ed. G. S. Rousseau and Roy Porter (Cambridge, U.K., 1980).

For information on any individual scientist, one should consult the *Dictionary of scientific biography,* ed. C. C. Gillispie, 16 vols. (New York, 1970–80). The articles are scholarly and are usually accompanied by extensive bibliographies.

I. The Character of the Enlightenment

Several older studies that are still of great value to the historian of science are Basil Willey, *The eighteenth century background, studies on the idea of nature in the thought of the period* (New York, 1940); Ernst Cassirer, *Philosophy of the Enlightenment,* trans. Fritz C. A. Koelln and James P. Pettegrove (Boston, 1951); and Preserved Smith, *A history of modern culture,* 2 vols. (New York, 1930), reissued in 1962 as *The Enlightenment, 1687–1776.* Willey's book is, as the title says, an investigation into the idea of nature during the Enlightenment; Cassirer's is more philosophical, with a neo-Kantian slant; and Smith's covers all of Enlightenment "culture," including the natural sciences. A modern study that I have used with great benefit is Peter Gay's *The Enlightenment, an interpretation,* 2 vols. (New York, 1969), especially vol. 2, "The science of freedom." Isaiah Berlin, *The age of Enlightenment* (New York, 1960), and George R. Havens, *The age of ideas, from reaction to revolution in eighteenth-century France* (New York, 1955), are good introductions.

Science and literature are the subject of Marjorie Hope Nicholson's *Newton demands the muse: Newton's "Opticks" and the eighteenth-century poets* (Princeton, 1946), and *Science and imagination* (Ithaca, 1956). The *éloges* at the Paris Academy of Sciences were important literary statements about science and scientists during the Enlightenment; their history is well documented in Charles B. Paul, *Science and immortality, the éloges of the Paris Academy of Sciences (1699–1791)* (Berkeley, 1980). I. B. Cohen's "The eighteenth-century origins of the concept of scientific revolution," *Journal of the History of Ideas* 37 (1976), 257–88, is the best source for information on this interesting topic.

Newton appeared in the Enlightenment as a heroic figure and also as a contributor to the solution of specific scientific problems. Historians have therefore judged his influence in many different ways. Pierre Brunet, *Introduction des théories de Newton en France au XVIIIe siècle avant 1738* (Paris, 1931) is limited to the debate in the Paris Academy of Sciences. More helpful are the studies by Henry Guerlac, *Newton on the Continent* (Ithaca, 1981); "Where the statue stood, divergent loyalties to Newton in the eighteenth century," in *Aspects of the eighteenth century,* ed. Earl R. Wasserman (Baltimore, 1965), pp. 317–34; and "Newton's changing reputation in the eighteenth century," in *Carl Becker's heavenly city revisited,* ed. Raymond O. Rockwood (Ithaca, 1958), 3–26. Other views are Gerd Buchdahl, *The image of Newton and Locke in the age of reason* (London, 1961); A. Rupert Hall, "Newton in France; a new view," *History of Science* 13 (1975), pp. 233–50; and Y. Elkana, "Newtonianism in the eighteenth century," *British Journal of the Philosophy of Science* 22 (1971), 297–306. Newtonian ideology as political radicalism is the theme of Margaret C. Jacob's *Newtonians and the English revolution, 1689–1720* (Ithaca, 1976), and *The radical Enlightenment; pantheists, Freemasons, and republicans* (London, 1981).

The dynamic aspect of Newton's philosophy and its importance for the eighteenth century are described in P. M. Heimann and J. E. McGuire, "Newtonian forces and Lockean powers: concepts of matter in eighteenth-century thought," *Historical Studies in the Physical Sciences* 3 (1971), 233–306; P. M. Heimann, "Newtonian natural philosophy and the scientific revolution," *History of Science* 11 (1973), 1–7; " 'Nature as a perpetual worker': Newton's aether and eighteenth-century natural philosophy," *Ambix* 20 (1973), 1–25; "Voluntarism and immanence: conceptions of nature in eighteenth-century thought," *Journal of the History of Ideas* 39 (1978), 271–83; and J. McEvoy and J. E. McGuire, "God and nature: Priestley's way of rational dissent," *Historical Studies in the Physical Sciences* 6 (1975), 325–404; and most recently Peter Harman, *Metaphysics and natural philosophy* (London, 1982).

Leibniz's role in the Enlightenment is equally controversial and less well understood. A general treatment is William Henry Barber, *Leibniz in France, from Arnauld to Voltaire: a study in French reactions to Leibnizianism, 1670–1760* (Oxford, 1955). Leibniz's disciple Christian Wolff was not an entirely accurate expositor of Leibniz's views. On Wolff, see Charles A. Corr, "Christian Wolff and Leibniz," *Journal of the History of Ideas* 36 (1975), 241–61, and Ronald S. Calinger, "The Newtonian-Wolffian confrontation in the St. Petersburg Academy of Sciences (1725–1746)," *Journal of World History* 11 (1968), 417–35, and "The Newtonian-Wolffian controversy (1740–1759)," *Journal of the History of Ideas* 30 (1969), 319–30.

II. Mathematics and the Exact Sciences

The history of mechanics has usually been written by physicists turned historians, and as a result it has tended to be a technical subject with a greater emphasis on mathematical manipulation than on the development of physical concepts. René Dugas's *History of mechanics*, trans. J. R. Maddox (Neuchatel, 1955), has material on the eighteenth century, as does István Szabó, *Geschichte der mechanischen Prinzipien und ihrer wichtigsten Anwendung* (Basel, 1977). The most penetrating investigation into the structure of eighteenth-century mechanics has been that of Clifford Truesdell. His article "A program toward rediscovering the rational mechanics of the age of reason," *Archive for History of Exact Science* 1 (1960–2),

1–36, emphasizes the formal mathematical character of mechanics during the Enlightenment. Others of his essays are in his *Six lectures on modern natural philosophy* (Berlin, 1966). But his most important writings are the lengthy introductions to volumes of Leonhard Euler's *Opera omnia* (Leipzig, 1912–; imprint varies) in which he analyzes the hundreds of books and papers published on mechanics during the Enlightenment. These introductions are "Rational fluid mechanics, 1687–1765," in *Opera omnia,* ser. 2, 12 (1954), i–cxxv; "I. The first three sections of Euler's treatise on fluid mechanics (1766)," "II. The theory of aerial sound (1687–1788)," and "III. Rational fluid mechanics (1765–1788)," *Opera omnia,* ser. 2, 13 (1956), vii–cxiii; and "The rational mechanics of flexible or elastic bodies, 1638–1799," in *Opera omnia,* ser. 2, 11, sec. 2 (1960).

On d'Alembert's mechanics, see my *Jean d'Alembert: science and the Enlightenment* (Oxford, 1970), and on the concepts of mechanics my articles "The influence of Malebranche on the science of mechanics during the eighteenth century," *Journal of the History of Ideas* 28 (1967), 193–210, and "The reception of Newton's second law of motion in the eighteenth century," *Archives internationales d'histoire des sciences* 78–9 (Jan.–June 1967), 43–65. Wilson L. Scott, "The significance of 'hard bodies' in the history of scientific thought," *Isis* 50 (1959), 199–210, treats the problem of discontinuous velocities; he greatly expands his discussion in *The conflict between atomism and conservation theory, 1644–1860* (London, 1970). J. R. Ravetz discusses other conceptual problems of mechanics in "The representation of physical quantities in eighteenth-century mathematical physics," *Isis* 52 (1961), 7–20, and "Vibrating strings and arbitrary functions" in *The logic of personal knowledge* (London, 1961), pp. 71–88. Other discussions of the concept of force are P. M. Heimann, " 'Geometry and nature': Leibniz and Johann Bernoulli's theory of motion," *Centaurus* 21 (1977), 1–26; Brian Ellis, "The existence of forces," *Studies in the History and Philosophy of Science* 7 (1976), 171–85; and Max Jammer, *Concepts of force: a study in the foundations of dynamics* (Cambridge, Mass., 1957). Various aspects of the *vis viva* controversy are described in my "Eighteenth-century attempts to resolve the *vis viva* controversy," *Isis* 56 (1965), 281–97; L. Laudan, "The *vis viva* controversy, a post mortem," *Isis* 59 (1968), 131–43; and P. M. Heimann and J. E. McGuire, "Cavendish and the *vis viva* controversy: a Leibnizian postscript," *Isis* 62 (1971), 225–7. Carolyn Iltis has placed the controversy in a wider context in her articles "The Leibnizian-Newtonian debates: natural philosophy and social psychology," *British Journal for the History of Science* 6 (1973), 343–77; "The decline of Cartesianism in mechanics: the Leibnizian-Cartesian debates," *Isis* 64 (1973), 356–73; and "Madame du Châtelet's metaphysics and mechanics," *Studies in the History and Philosophy of Science* 8 (1977), 29–48. Erwin N. Hiebert treats the same subject as background to energy conservation in his *Historical roots of the principle of conservation of energy* (Madison, 1962).

Historians of mathematics have tended to work more in the seventeenth and nineteenth centuries than in the eighteenth, but that is beginning to change. A good survey arranged topically is Morris Kline, *Mathematical thought from ancient to modern times* (New York, 1972). More specific studies are I. Grattan-Guinness, *The development of the foundations of mathematical analysis from Euler to Riemann* (Cambridge, Mass., 1970), which contains a chapter on the eighteenth century, and Herman H. Goldstine, *A history of the calculus of variations from the seventeenth through the nineteenth century* (New York, 1980). H. J. M. Bos makes a conscious effort to avoid anachronistic ideas and notations in his "Differentials, higher order differentials and the derivative in the Leibnizian calculus," *Archive for History of Exact Science* 14 (1974), 1–90. For Lagrange's work, see Robin Rider Hamburg, "The theory of equations in the eighteenth century: the work of Joseph Lagrange," *Archive for History of Exact Science* 16 (1976), 17–36. Also, on the connection between geodesy and recent advances in mathematics in the 1730s, see John Greenberg, "Geodesy in Paris in the 1730s and the Paduan connection," *Historical Studies in the Physical Sciences* 13 (1983), 239–60, and "Alexis Fontaine's integration of ordinary differential equations and the origins of the calculus of several variables," *Annals of Science* 39 (1982), 1–36.

The role of Malebranche in the spread of calculus on the Continent is described by Henry Guerlac in his *Newton on the continent* (Ithaca, 1981) and by André Robinet, "Le groupe malebranchiste, introducteur du calcul infinitesimal en France," *Revue d'histoire des sciences* 13 (1960), 287–308. The most detailed biography of Maupertuis is still Pierre Brunet's *Maupertuis* 2 vols. (Paris, 1929). For more information on Maupertuis, see Harcourt Brown, "Maupertuis philosophe: enlightenment and the Berlin academy," *Studies on Voltaire and the Eighteenth Century* 24 (1963), 255–69; "From London to Lapland: Maupertuis, Johann Bernoulli I, and *La terre applatie,* 1728–1738," in Charles G. S. Williams, ed., *Literature and history in the age of ideas* (Columbus, 1975), pp. 69–96. A highly entertaining account of the voyage to Lapland is given in Tom R. Jones, *The figure of the earth* (Lawrence, 1967). Harry Woolf, *The transits of Venus* (New York, 1959), describes a later internationally organized astronomical expedition. Accounts of the debate over the motion of the lunar apogee are Philip Chandler, "Clairaut's critique of Newtonian attraction: some insights into his philosophy of science," *Annals of Science* 32 (1975), 369–78, and Craig B. Waff, "Alexis Clairaut and his proposed modification of Newton's inverse-square law of gravitation," in *Avant, avec, après Copernic* (Paris, 1975), pp. 281–8.

There are few biographies of eighteenth-century mathematicians. Charles C. Gillispie, *Lazare Carnot, savant* (Princeton, 1971), describes Carnot's mechanics, and Roger Hahn is preparing a biography of Laplace that he describes in *Laplace as a Newtonian scientist,* Williams Andrews Clark Memorial Library (Los Angeles, 1967).

Astronomy is a largely uncharted subject for the eighteenth century. A very thorough (and technical) treatment is Curtis A. Wilson's "Perturbations and solar tables from Lacaille to Delambre, the rapprochement of observation and theory," *Archive for History of Exact Science* 22 (1980), 53–304. Antonie Pannekoek, *A history of astronomy* (New York, 1961), contains a good general treatment of eighteenth-century astronomy, and E. J. Aiton, *The vortex theory of planetary motions* (London, 1972), discusses theories of planetary motion. Eric G. Forbes wrote many articles on Tobias Mayer; a good beginning is "Tobias Mayer's contributions to observational astronomy," *Journal of the History of Astronomy* 11 (1980), 28–49.

III. Experimental Physics

The history of physics has tended to be extremely anachronistic, probably because the earliest histories were written by nineteenth-century German physicists for the university physics curriculum. The best of the early German histories is Ferdinand Rosenberger, *Die Geschichte der Physik,* 3 vols. (Braunschweig, 1882–90), an encyclopedic treatment of all physics, but as seen from the end of the nineteenth century. John L. Heilbron, *Electricity in the seventeenth and eighteenth centuries: a study of early modern physics* (Berkeley, 1979), describes the emergence of experimental physics in its cultural context, with special attention to the history of electricity. Heilbron documents the rise of the profession in the universities, in the academies, and in demonstration lectures. His description of electrical experiments is detailed, and the book contains an extensive bibliography. Hans Schimank, "Die Wandlung des Begriffs 'Physik' während der ersten Halfte des 18. Jahrhunderts," in K. H. Manegold, ed., *Wissenschaft, Wirtschaft und Technik* (Munich, 1969), pp. 454–68, describes the emergence of "physics" understood in its modern sense.

An earlier history of electricity with special emphasis on Benjamin Franklin and his assimilation of Newtonian philosophy is I. Bernard Cohen, *Franklin and Newton* (Philadelphia, 1956). An article on a more particular subject by the same author is "Prejudice against the introduction of lightning rods," *Journal of the Franklin Institute* 253 (1952), 393–440. The important work of Aepinus has been described by Roderick W. Home in "Aepinus and the English electricians: the dissemination of a scientific theory," *Isis* 63 (1972), 190–204, and "Aepinus, the tourmaline crystal, and the theory of electricity and magnetism," *Isis* 67 (1976), 21–30, and on magnetism see his "Newtonianism and the theory of the magnet," *History of Science* 15 (1977), 252–66.

New measuring instruments and demonstration apparatus were important for the creation of experimental physics. Heilbron describes electrical apparatus; Maurice Daumas treats the whole range of instrumentation in his *Les instruments scientifiques aux XVIIe et XVIIIe siècles* (Paris, 1953) and "Precision of measurement and physical and chemical research in the eighteenth century," in Alistair C. Crombie, ed., *Scientific change* (New York, 1963), pp. 418–30. C. Stewart Gillmor gives an account of Coulomb's work in *Coulomb and the evolution of physics and engineering in eighteenth-century France* (Princeton, 1971), and on Cavendish see Russell K. McCormmach, "Henry Cavendish: a study of rational empiricism in eighteenth-century natural philosophy," *Isis* 60 (1969), 293–306. W. E. Knowles Middleton describes meteorologic instruments in *The history of the barometer* (Baltimore, 1964), *A history of the thermometer and its use in meteorology* (Baltimore, 1966), and *Invention of the meteorological instruments* (Baltimore, 1969).

The rise of the subtle fluids in the eighteenth century and their coincidence with a recognition of Newton's ether theories has led Robert E. Schofield to suggest a major philosophical division between those physicists who accepted an ether and those who wished to explain everything by action at a distance. He develops these ideas in *Mechanism and materialism: British natural philosophy in an age of reason* (Princeton, 1970). More recently G. N. Cantor and M. J. S. Hodge have edited a volume of articles on the ether: *Conception of ether: studies in the history of ether theories, 1740–1900* (Cambridge, U.K., 1981).

A good introduction to theories of heat is Douglas McKie and Niels H. de V. Heathcote, *The discovery of specific and latent heats* (London, 1935). Also valuable is *Partners of science: letters of James Watt and Joseph Black,* ed. Eric Robinson and Douglas McKie (London, 1970). The notion of "fire" in the eighteenth century is still a puzzle, but see Hélène Metzger, "La théorie du feu d'après Boerhaave," *Revue philosophique* 109 (1930), 253–85. Henry Guerlac describes the important calorimetry experiments of Lavoisier and Laplace in "Chemistry as a branch of physics: Laplace's collaboration with Lavoisier," *Historical Studies in the Physical Sciences* 7 (1976), 193–276. A good biography of Rumford is Sanborn C. Brown's *Benjamin Thompson, Count Rumford* (Cambridge, Mass., 1979).

For physics in Holland, see Pierre Brunet, *Les physiciens hollandais et la méthode expérimentale en France au XVIIIe siècle* (Paris, 1926), and E. G. Ruestow, *Physics at seventeenth and eighteenth century Leiden* (The Hague, 1973). For Scotland, see Richard Olson, *Scottish philosophy and British physics, 1750–1880: a study in the foundations of the Victorian scientific style* (Princeton, 1975), and John R. R. Christie, "The origins and development of the Scottish scientific community, 1680–1760," *History of Science* 12 (1974), 122–41. For Russia, see Valentin Boss, *Newton and Russia: the early influence, 1698–1796* (Cambridge, Mass., 1972).

IV. Chemistry

The history of the Chemical Revolution has itself undergone a revolution in recent years. My interpretation is closest to that of J. B. Gough, "The origins of Lavoisier's theory of the gaseous state," in Harry Woolf, ed., *The analytic spirit: essays in the history of science in honor of Henry Guerlac* (Ithaca, 1981), pp. 15–39. It also draws from the many essays of Henry Guerlac. These include two books on Lavoisier: *Lavoisier – the crucial year: the background and origin of his first experiments on combustion in 1772* (Ithaca, 1961) and *Antoine-Laurent Lavoisier, chemist and revolutionary* (New York, 1975). His many articles on the history of chemistry are now conveniently collected in his *Essays and papers in the history of modern science* (Baltimore, 1977). Douglas McKie's *Antoine Lavoisier, scientist, economist, social reformer* (New York, 1962) describes Lavoisier's multitudinous activities. Additional material on Lavoisier's theory of heat and the gaseous state is in Robert Siegfried, "Lavoisier's view of the gaseous state and its early application to pneumatic chemistry," *Isis* 63 (1972), 59–78, and Robert J. Morris, "Lavoisier and the caloric theory," *British Journal for the History of Science* 6 (1972), 1–38, and Robert Fox, *The caloric theory of gases: from Lavoisier to Regnault* (Oxford, 1971).

Arthur L. Donovan describes the work of the Scottish chemists in *Philosophical chemistry in the Scottish Enlightenment: the doctrines and discoveries of William Cullen and Joseph Black* (Edinburgh, 1975), and "Pneumatic chemistry and Newtonian natural philosophy in the eighteenth century: William Cullen and Joseph Black," *Isis* 67 (1976), 217–28. The mechanical philosophy in chemistry is the subject of Arnold W. Thackray, *Atoms and powers: an essay on Newtonian matter-theory and the development of chemistry* (Cambridge, Mass., 1970). On the importance of Stahl for the chemical revolution, see Hélène Metzger, *Newton, Stahl, Boerhaave et la doctrine chimique* (Paris, 1930), and Rhoda Rappaport, "Rouelle and Stahl – the phlogistic revolution in France," *Chymia* 7 (1961), 73–102. Other more general interpretations are Maurice P. Crosland, *Historical studies in the language of chemistry* (London, 1962), and Crosland, "The development of chemistry in the eighteenth century," *Studies on Voltaire and the Eighteenth Century* 24 (1963), 369–441.

Attempts to solve the puzzle of how Dalton came to his atomic theory are Henry Guerlac, "The background to Dalton's atomic theory," in Donald S. L. Cardwell, ed., *John Dalton and the progress of science* (Manchester, U.K., 1968), pp. 57–91; Arnold Thackray, *John Dalton: critical assessments of his life and science* (Cambridge, Mass., 1972), and "The origin of Dalton's chemical atomic theory: Daltonian doubts resolved," *Isis* (1966), 44–5.

V. Natural History and Physiology

The most thorough study of eighteenth-century theories of generation and their importance for natural history and for Enlightenment philosophy is Jacques Roger, *Les sciences de la vie dans la pensée française du XVIIIe siècle* (Paris, 1963). Unfortunately it has not been translated into English. General histories of biology that have chapters on the eighteenth century are Erik Nordenskiöld, *The history of biology: a survey* (New York, 1928), and Lois N. Magner, *History of the life sciences* (New York, 1979). P. C. Ritterbush, *Overtures to biology: the speculations of eighteenth-century naturalists* (New Haven, 1964), discusses several special problems, including electricity and life. Physics, chemistry, and life are the subjects of Everett Mendelsohn, *Heat and life, the development of the theory of animal heat* (Cambridge, Mass., 1964); Leonard K. Nash, *Plants and the atmosphere,* in Harvard Case Histories in Experimental Science, no. 5 (Cambridge, Mass., 1952); and Jacques Roger, "Chimie et biologie des 'molécules organiques' de Buffon à la 'physico-chimie' de Lamarck," *History and Philosophy of the Life Sciences* 1 (1979), 43–64.

Theodore M. Brown discusses the changes in English physiology in "From mechanism to vitalism in eighteenth-century English physiology," *Journal of the History of Medicine* 7 (1974), 179–216, and Thomas S. Hall covers the entire subject in *Ideas of life and matter: studies in the history of general physiology, 600 BC – 1900 AD,* 2 vols. (Chicago, 1969). Lester S. King treats the medical dimension in *The philosophy of medicine: the early eighteenth century* (Cambridge, Mass., 1978); "Theory and practice in eighteenth-century medicine," *Studies in Voltaire and the Eighteenth Century,* 153 (1976), 1201–18, and "Stahl and Hoffman: a study in eighteenth-century animism," *Journal of the History of Medicine* 19 (1964), 118–30.

A good introduction to the history of theories of generation is Elizabeth B. Gasking, *Investigations into generation, 1651–1828* (Baltimore, 1967). Also valuable is Charles W. Bodemer, "Regeneration and the decline of preformationism in eighteenth-century embryology," *Bulletin of the History of Medicine* 38 (1964), 20–31. For the philosophical implications of regeneration, see A. Vartanian, "Trembley's polyp, La Mettrie, and eighteenth-century French materialism," in P. P. Wiener and A. Noland, eds., *Roots of scientific thought* (New York, 1957), pp. 497–516. Iris Sandler, "The reexamination of Spallanzani's interpretation of the role of the spermatic animalcules in fertilization," *Journal of the History of Biology* 6 (1973), 193–223, is the basis for my account of Spallanzani. *Forerunners of Darwin: 1745–1859,* ed. Bentley Glass, Oswei Temkin, and William L. Strauss, Jr. (Baltimore, 1959), contains articles on a variety of subjects in eighteenth-century life sciences.

Shirley A. Roe, *Matter, life, and generation: eighteenth-century embryology and the Haller–Wolff debate* (Cambridge, U.K., 1981) extends the work of her earlier articles on Haller and Caspar Wolff, "The development of Albrecht von Haller's views on embryology," *Journal of the History of Biology* 8 (1975), 167–90, and "Rationalism and embryology: Caspar Friedrich Wolff's theory of epigenesis," *Journal of the History of Biology* 12 (1979), 1–43. John Farley, *The spontaneous generation controversy from Descartes to Oparin* (Baltimore, 1977) has material on the eighteenth century. On Bordeu, see Elizabeth L. Haigh, "Vitalism, the soul and sensibility: the physiology of Théophile Bordeu," *Journal of the History of Medicine* 31 (1976), 3–41, and Herbert Dieckmann, "Théophile Bordeu und Diderot's Rève de d'Alembert," *Romanische Forschungen* 3 (1938), 55–122. On Réaumur, see Jean Torlais, *Réaumur, un esprit encyclopédique en dehors de l'Encyclopédie* (Paris, 1937).

Natural history and taxonomy are the most complicated of all issues in the life sciences of the eighteenth century. Three articles by Phillip R. Sloan reveal many of the points at issue: "John Locke, John Ray, and the problem of the natural system," *Journal of the*

History of Biology 5 (1972), 1–53; "The Buffon-Linnaeus controversy," *Isis* 67 (1976), 356–75; and "Buffon, German biology, and the historical interpretation of biological species," *British Journal for the History of Science* 12 (1979), 109–53. On Buffon, see John Lyon and Phillip R. Sloan, eds. and trans., *From natural history to the history of nature: readings from Buffon and his critics* (Notre Dame, 1981); Peter J. Bowler, "Bonnet and Buffon: theories of generation and the problem of species," *Journal of the History of Biology* 6 (1973), 259–81; and Paul L. Farber, "Buffon and the concept of species," *Journal of the History of Biology* 5 (1972), 259–84. Jacques Roger, "Diderot et Buffon en 1749," *Diderot Studies* 4 (1963), 221–36, describes Buffon's early influence on Diderot.

The classic study of the great chain of being is Arthur O. Lovejoy, *The great chain of being: a study in the history of an idea* (Cambridge, Mass., 1936). Bonnet's use of the chain is described in Lorin Anderson, "Charles Bonnet's taxonomy and chain of being," *Journal of the History of Ideas* 37 (1976), 45–58. The emergence of the idea of an ecological balance is discussed in Frank N. Egerton, "Changing concepts of the balance of nature," *Quarterly Review of Biology* 48 (1973), 322–50, and for Linnaeus's view see *L'équilibre de la nature,* trans. Bernard Jasmin with introduction by Camille Limoges (Paris, 1972). Recent interest in Linnaeus has produced several important studies: James L. Larson, *Reason and experience: the representation of natural order in the work of Carl von Linné* (Berkeley, 1971), and "The species concept of Linnaeus," *Isis* 59 (1968), 281–99; and Frans A. Stafleu, *Linnaeus and the Linnaeans* (Utrecht, 1971). On the relationship between the historical view and the evolutionary view of nature, see Wolf Lepenies, "De l'histoire naturelle à l'histoire de la nature," *Dix-huitième siècle* 11 (1979), 179–81, and Peter J. Bowler, "Evolutionism in the Enlightenment," *History of Science* 12 (1974), 159–83.

Two books covering the geology of the eighteenth century are Roy Porter, *The making of geology: earth science in Britain, 1660–1815* (Cambridge, U.K., 1977), and Gordon L. Davies, *The earth in decay: a history of British geomorphology, 1578–1878* (New York, 1969). Different views of Hutton and Werner are P. Gerstner, "James Hutton's theory of the earth and his theory of matter," *Isis* 59 (1968), 26–31; Martin J. S. Rudwick, "Hutton and Werner compared: George Greenough's geological tour of Scotland in 1805," *British Journal for the History of Science* 1 (1962), 117–35; V. A. Eyles, "Abraham Gottlob Werner (1749–1817) and his position in the history of the mineralogical and geological sciences," *History of Science* 3 (1964), 102–15; and Alexander Ospovat, "The distortion of Werner in Lyell's Principles of Geology," *British Journal for the History of Science* 9 (1976), 190–98. There is also a very good discussion of Hutton and Werner in the first chapter of Mott T. Greene, *Geology in the nineteenth century: changing views of a changing world* (Ithaca, 1982). And finally, for a discussion of questions that have not been treated to date, see Joseph Schiller, "Queries, answers and unsolved problems in eighteenth-century biology," *History of Science* 12 (1974), 184–99.

VI. The Moral Sciences

The literature on the intellectual history of the Enlightenment is so vast that an essay of this size can only mention a few titles. The two volumes of Peter Gay's *The Enlightenment: an interpretation* (see the section in this bibliographic essay on "The character of the Enlightenment") both contain long bibliographic essays, which provide a good starting place. Two books that discuss the meaning of "reason" in the Enlightenment are Charles Frankel, *The faith of reason: the idea of progress in the French Enlightenment* (New York, 1948), and Robert McRae, *The problem of the unity of the sciences: Bacon to Kant* (Toronto, 1961). The most provocative and also the most readable essay that I have read on the Enlightenment is Carl Becker's *The heavenly city of the eighteenth-century philosophers* (New Haven, 1935); it has been ably judged and analyzed in Raymond O. Rockwood, ed., *Carl Becker's Heavenly City revisited,* (Ithaca, 1958). An old but still useful book on political theory is Kingsley Martin, *French liberal thought in the eighteenth century: a study of political ideas from Bayle to Condorcet* (Boston, 1929). More difficult is Ira O. Wade, *The structure and form of the French Enlightenment,* 2 vols. (Princeton, 1977).

Michel Foucault has completely restructured the Enlightenment in his *The order of things: an archaeology of the human sciences* (New York, 1970), and *Madness and civilization: a history of insanity in the age of reason* (New York, 1965). These books contain many insights, but the insights are not obtained without effort by the reader. Ernest C. Mossner, *The life of David Hume* (Edinburgh, 1954), is a good biography. D'Alembert's campaign for philosophy is described in Ronald Grimsley, *Jean d'Alembert, 1717–83* (Oxford, 1963), and John N. Pappas, *Voltaire and d'Alembert* (Bloomington, 1962). John Lough provides an introduction to the *Encyclopédie* in his *The Encyclopédie* (London, 1971), and *The contributors to the "Encyclopédie"* (London, 1973).

A very complete account of the French Academy of Sciences and its demise during the French Revolution is Roger Hahn's *Anatomy of a scientific institution: the Paris Academy of Sciences, 1666–1803* (Berkeley, 1971). René Taton, ed., *Enseignement et diffusion des sciences en France au XVIIIe siècle* (Paris, 1964), contains important articles on scientific institutions, and the entire subject of government policy towards science in France has now been studied in detail by C. C. Gillispie, *Science and polity in France at the end of the old regime* (Princeton, 1980). Another aspect of science (or pseudoscience) and radical politics is Robert Darnton's *Mesmerism and the end of the Enlightenment in France* (Cambridge, Mass., 1968). A very complete biography of Diderot, the work of a lifetime, is Arthur Wilson's *Diderot* (New York, 1972). Other interpretations of Diderot are Herbert Dieckmann's "The influence of Francis Bacon on Diderot's *Interprétation de la nature*," *Romanic Review* 34 (1943), 303–30, and Aram Vartanian's *Diderot and Descartes: a study of scientific naturalism in the Enlightenment* (Princeton, 1953).

The best description of the attempts by the philosophes to create a social science is Keith Michael Baker's *Condorcet: from natural philosophy to social mathematics* (Chicago, 1975). This book contains a wealth of information on many aspects of the Enlightenment, not just Condorcet. Frank E. Manuel, *The prophets of Paris: Turgot, Condorcet, Saint-Simon, Fourier and Comte* (New York, 1965), carries the history of social theory on into the nineteenth century. Lorraine J. Daston describes the interaction between social theory and probability theory in "Probabilistic expectation and rationality in classical probability theory," *Historia mathematica* 7 (1980), 234–60, and "D'Alembert's critique of probability theory," *Historia mathematica* 6 (1979), 259–79. More general works on probability theory are Ian Hacking, *The emergence of probability: a philosophical study of early ideas about probability, induction and statistical inference* (London, 1975); L. E. Maĭstrov, *Probability theory: a historical sketch*, trans. S. Kotz (New York, 1974); and Isaac Todhunter, *A history of the mathematical theory of probability from the time of Pascal to that of Laplace* (Cambridge, U.K., 1865).

Sources of Quotations

Note: I have quoted from original sources wherever possible, if the original is in English. Where the original is not in English, I have sought a suitable English-language source. All translations from original sources are my own.

I. The Character of the Enlightenment

1 D'Alembert, quoted in Ernst Cassirer, *The philosophy of the Enlightenment,* trans. Fritz C. A. Koelln and James P. Pettegrove (Boston, 1955), pp. 3–4.
2 D'Alembert, *Encyclopédie* article "Expérimentale," quoted in Peter Gay, *The Enlightenment: an interpretation,* 2 vols. (New York, 1966), vol. I, *The rise of modern paganism,* p. 319.
3 See I. Bernard Cohen, "The eighteenth-century origins of the concept of scientific revolution," *Journal of the History of Ideas* 37 (1976), 267.
4 Ibid., p. 263.
5 Kant, quoted in Gay, *The Enlightenment,* I, 20.
6 Locke, quoted in Basil Willey, *The eighteenth-century background: studies on the idea of nature in the thought of the period* (New York, 1940), p. 7.
7 Robert Boyle, "A disquisition about the final causes of natural things" (1688), in *Works,* 5 vols. (London, 1744), IV, 523.
8 Isaac Newton, *Opticks,* 4th ed. (1730; rpt. New York, 1952), p. 405.
9 Rohalt, quoted in John L. Heilbron, *Electricity in the seventeenth and eighteenth centuries; a study of early modern physics* (Berkeley, 1979), p. 11.

II. Mathematics and the Exact Sciences

1 Lagrange to d'Alembert, September 21, 1781, quoted in Thomas L. Hankins, *Jean d'Alembert: science and the Enlightenment* (Oxford, 1970), p. 99.
2 Lacroix, quoted in Morris Kline, *Mathematical thought from ancient to modern times* (New York, 1972), p. 623.

3 Isaac Barrow, *Mathematical lectures read in the publick schools at the University of Cambridge (1664–5)*, trans. John Kirby (London, 1734), p. 28.
4 Isaac Newton, *Opticks*, 4th ed. (1730; rpt. New York, 1952), p. 404.
5 Voltaire, quoted in Hankins, *D'Alembert*, p. 115.
6 Lagrange, quoted in Hankins, *D'Alembert*, p. 232.
7 Jean d'Alembert, "Elémens de philosophie," in *Mélanges de littérature, d'histoire et de philosophie*, 5 vols. (Amsterdam, 1770), IV, 231.
8 D'Alembert, quoted in Hankins, *D'Alembert*, p. 39.

III. Experimental Physics

1 Jean d'Alembert, *Preliminary discourse to the Encyclopedia*, trans. Richard N. Schwab (Indianapolis, 1963), p. 22.
2 Benjamin Franklin, *Benjamin Franklin's experiments, a new edition of "Experiments and observations on electricity" (1747)*, ed. I. Bernard Cohen (Cambridge, Mass., 1941), p. 174.
3 Ibid., pp. 174–5.
4 Musschenbroek, quoted in J. L. Heilbron, *Electricity in the seventeenth and eighteenth centuries: a study of early modern physics* (Berkeley, 1979), p. 313.
5 Nollet, quoted in Heilbron, *Electricity*, p. 318.
6 Joseph Black, *Lectures on the elements of chemistry . . . by the late Joseph Black, M.D.*, ed. John Robinson, 3 vols. (Philadelphia, 1806–7), I, 112.

IV. Chemistry

1 Turgot, quoted in J. B. Gough, "The origin of Lavoisier's theory of the gaseous state," in *The analytic spirit: essays in the history of science*, ed. Harry Woolf (Ithaca, 1981), pp. 29–30.
2 Robert Hooke, *Micrographia* (London, 1665), p. 103.
3 Lavoisier, quoted in Henry Guerlac, *Essays and papers in the history of modern science* (Baltimore, 1977), p. 394 (my translation).
4 Lavoisier, quoted in Douglas McKie, *Antoine Lavoisier, scientist, economist, social reformer* (New York, 1962), p. 74.
5 Ibid., p. 88.
6 Joseph Priestley to Sir John Pringle, March 15, 1775, published as "An account of further discoveries in air," Royal Society of London *Philosophical Transactions* 65 (1775), 388.
7 Priestley, quoted in James Bryant Conant, *The Overthrow of the phlogiston theory: the chemical revolution of 1775–1789*, Harvard Case Histories in Experimental Science, no. 2 (Cambridge, Mass., 1948), p. 34.
8 Lavoisier, quoted in René Fric, "Contribution à l'étude de l'évolution des idées de Lavoisier sur la nature de l'air et sur la calcination des métaux," *Archives internationales d'histoire des sciences* 12 (1959), 151 (my translation).
9 Henry Cavendish, "Experiments on air," Royal Society of London *Philosophical Transactions* 74 (1784), 129.
10 Lavoisier, quoted in McKie, *Lavoisier*, p. 191.

11 Lavoisier, quoted in H. S. van Klooster, "Franklin and Lavoisier," *Journal of Chemical Education* 23 (1946), 107–9.

V. Natural History and Physiology

1 Parsons, quoted in Robert E. Schofield, *Mechanism and materialism: British natural philosophy in an age of reason* (Princeton, 1969), p. 195.
2 Diderot, *Rêve de d'Alembert,* ed. Paul Vernière (Paris, 1951), pp. 69–71.
3 Diderot to Duclos, October 10, 1765, in Diderot, *Oeuvres philosophiques,* ed. Paul Vernière (Paris, 1961), p. 249.
4 Köhlreuter, quoted in Robert C. Olby, *Origins of Mendelism* (New York, 1966), p. 27.
5 Caspar Wolff, quoted in Elizabeth B. Gasking, *Investigations into generation, 1651–1828* (Baltimore, 1967), p. 104.
6 Bonnet, quoted in Gasking, *Investigations into generation,* p. 124.
7 Adanson, quoted in Phillip R. Sloan, "John Locke, John Ray, and the problem of the natural system," *Journal of the History of Biology* 5 (1972), 3.
8 Linnaeus, quoted in Sloan, "John Locke, John Ray," p. 4.
9 Buffon, "Premier discours de la manière d'étudier et de traiter l'histoire naturelle," quoted in Phillip R. Sloan, "The Buffon-Linnaeus controversy," *Isis* 67 (1976), 359.
10 Ibid.
11 J. L. Lignac, *Lettres à un Amériquain sur l'histoire naturelle . . . de Buffon,* 2d ed. (Paris, 1756), quoted in Sloan, "Buffon-Linnaeus," p. 361.
12 Buffon, quoted in Sloan, "Buffon-Linnaeus," p. 370.
13 Hutton, quoted in Mott T. Greene, *Geology in the nineteenth century: changing views of a changing world* (Ithaca, 1982), p. 25.
14 Werner, quoted in Greene, *Geology,* p. 33.

VI. The Moral Sciences

1 The quotations by Turgot in this and the previous paragraph are from Keith M. Baker, *Condorcet: from natural philosophy to social mathematics* (Chicago, 1975), pp. 203–7.
2 David Hume, *A treatise of human nature* (1739–40), ed. L. A. Selby-Bigge (Oxford, 1888), p. xxiii.
3 The quotations by Montesquieu in this and the previous paragraph are from Montesquieu, *Spirit of the laws,* ed. David Wallace Carrithers, trans. Thomas Nugent (Berkeley, 1977), p. 98, 244–46.
4 D'Alembert, quoted in Thomas L. Hankins, *Jean d'Alembert: science and the Enlightenment* (Oxford, 1970), p. 81.
5 Voltaire, quoted in Peter Gay, *The Enlightenment: an interpretation,* 2 vols. (New York, 1966), vol. II, *The science of freedom,* p. 324.
6 D'Alembert to Samuel Formey, September 19, 1749, quoted in Hankins, *D'Alembert,* p. 68.
7 Attorney-general Omer Joly de Fleury to the Parlement of Paris, January 1759, quoted in Hankins, *D'Alembert,* p. 72.

8 Jean d'Alembert, *Preliminary discourse to the Encyclopedia,* trans. Richard N. Schwab (Indianapolis, 1963), p. 121.
9 D'Alembert, quoted in Hankins, *D'Alembert,* p. 76.
10 Diderot, quoted in Hankins, *D'Alembert,* p. 100.
11 D'Alembert, quoted in Hankins, *D'Alembert,* p. 101.
12 Voltaire, quoted in Hankins, *D'Alembert,* p. 102.
13 Article 31 of the 1699 regulations of the Paris Academy of Sciences, quoted in Roger Hahn, *The anatomy of a scientific institution; the Paris Academy of Sciences, 1666–1803* (Berkeley, 1971), p. 23.
14 Rousseau, quoted in Gay, *The Enlightenment,* II, 536–8.
15 Voltaire, quoted in George R. Havens, *The age of ideas: from reaction to revolution in eighteenth-century France* (New York, 1955), pp. 257–8.
16 Voltaire, quoted in Havens, *Age of ideas,* p. 257.
17 Hume, *Treatise,* p. 415.
18 D'Holbach, quoted in Gay, *The Enlightenment,* II, 194.
19 Diderot, *Rameau's nephew and other works,* trans. Jacques Barzun and Ralph H. Bowers (Garden City, N. Y., 1956), pp. 209–11.
20 Locke, *An essay concerning human understanding,* ed. Alexander Campbell Fraser, 2 vols. (Oxford, 1894), II, 217–18, 360.
21 Huygens, quoted in Lorraine J. Daston, "Probabilistic expectations and rationality in classical probability theory," *Historia Mathematica* 7 (1980), 236.
22 Laplace, quoted in Daston, "Probabilistic expectations," p. 242.
23 Beccaria, quoted in Gay, *The Enlightenment,* II, 443.
24 Condorcet, quoted in Baker, *Condorcet,* p. 86.
25 D'Alembert, quoted in Hankins, *D'Alembert,* p. 146.
26 Condorcet, quoted in Baker, *Condorcet,* p. 81. The quotations from Condorcet that follow are from pp. 169, 236, 274, 324, 329, and 350.
27 Abraham de Moivre, *The doctrine of chances, or a method of calculating the probability of events in play* (London, 1818). The quotation is from the dedication to Isaac Newton.

Index